String Theory
FOR
DUMMIES®

by Andrew Zimmerman Jones
with Daniel Robbins, PhD in Physics

WILEY

Wiley Publishing, Inc.

About the Author

Andrew Zimmerman Jones is the Physics Guide at About.com, a New York Times Company, where he writes and edits news and articles on all areas of physics. He spends his days working as an editor for an educational assessment company. He holds a bachelor's degree in physics from Wabash College, where he also studied mathematics and philosophy, and a master's degree in mathematical education from Purdue University.

In addition to work for About.com, Andrew has written a number of nonfiction essays and reviews, which have appeared in *The Internet Review of Science Fiction, EpicSFF.com, Pink Floyd and Philosophy, Black Gate,* and *Heroes and Philosophy.* His fiction credits include short stories in *Abyss and Apex, KidVisions, The Four Bubbas of the Apocalypse,* and *International House of Bubbas.*

He has been a member of Mensa since the eighth grade and has been intensely interested in both science and science fiction since even earlier. Along the way, he's also become an Eagle Scout, a Master Mason in the Freemasons, and won the Harold Q. Fuller Prize in Physics at Wabash College. His plan for world domination nears completion with the publication of this book.

Andrew lives in central Indiana with his beautiful wife, Amber, and son, Elijah. When he's not writing or editing, he is most often found reading, playing games, watching television, investigating bizarre scientific phenomena, or updating his personal Web page, which can be found at www.azjones.info. Andrew also regularly reports on any new string theory implications on his Web site at physics.about.com.

Dedication

This book is dedicated to my loving and lovely wife, Amber Eckert-Jones. While physicists still search for a law to unify all of the forces in the physical universe, I don't need to, because all the forces in my universe come together in you.

Author's Acknowledgments

I must first profoundly thank my agent, Barb Doyen, for approaching me with this project. My deepest thanks and appreciations go out to the wonderful editorial staff at Wiley: Alissa Schwipps for her valuable input at every step in the process, Vicki Adang for her ability to turn my scientific babble into coherent explanations, and Stacy Kennedy for gathering together such a great team in the first place. I also very much appreciated the constructive and at times critical input of Dr. Rolf Schimmrigk of Indiana University, South Bend, who provided initial technical editing on the book. In addition, I'm profoundly thankful for the extremely detailed technical expertise, review, and frequent discussions offered by Dr. Daniel Robbins of the Weinberg Theory Group at the University of Texas at Austin.

Without the wonderful staff at About.com, notably the Education Channel editor Madeleine Burry, I would never have had the opportunity to grow as a writer in this field. Also to author Robert J. Sawyer, for his mentorship and friendship over the years. Thanks to you all!

Many thanks to physicists Lee Smolin and John W. Moffat of the Perimeter Institute for Theoretical Physics, Leonard Susskind of Stanford University, and Sylvester James Gates, director of the University of Maryland's Center for String and Particle Theory, for e-mail exchanges that have helped to clarify various points throughout the writing of this book.

Finally, my thanks go out to my wife, Amber, and son, Elijah, for putting up with me, even when I was driven frantic by deadlines. Thanks also to my mother, Nancy Zimmerman, and mother-in-law, Tina Lewis, for their help in keeping the family entertained while I worked feverishly in the basement.

Publisher's Acknowledgments

We're proud of this book; please send us your comments through our Dummies online registration form located at `http://dummies.custhelp.com`. For other comments, please contact our Customer Care Department within the U.S. at 877-762-2974, outside the U.S. at 317-572-3993, or fax 317-572-4002.

Some of the people who helped bring this book to market include the following:

Acquisitions, Editorial, and Media Development

Senior Project Editor: Alissa Schwipps

Acquisitions Editor: Stacy Kennedy

Senior Copy Editor: Victoria M. Adang

Assistant Editor: Erin Calligan Mooney

Editorial Program Coordinator: Joe Niesen

Technical Editor: Rolf Schimmrigk

Senior Editorial Manager: Jennifer Ehrlich

Editorial Assistants: Jennette ElNaggar, David Lutton

Art Coordinator: Alicia B. South

Cover Photo: © Laguna Design / Photo Researchers, Inc.

Cartoons: Rich Tennant (`www.the5thwave.com`)

Composition Services

Project Coordinator: Lynsey Stanford

Layout and Graphics: Ana Carrillo, Melissa K. Jester, Christin Swinford, Christine Williams

Special Art: Precision Graphics

Proofreaders: Melissa Cossell, Sossity R. Smith

Indexer: Infodex Indexing Services, Inc.

Special Help

Matthew Headrick

Publishing and Editorial for Consumer Dummies

   Diane Graves Steele, Vice President and Publisher, Consumer Dummies

   Kristin Ferguson-Wagstaffe, Product Development Director, Consumer Dummies

   Ensley Eikenburg, Associate Publisher, Travel

   Kelly Regan, Editorial Director, Travel

Publishing for Technology Dummies

   Andy Cummings, Vice President and Publisher, Dummies Technology/General User

Composition Services

   Debbie Stailey, Director of Composition Services

Contents at a Glance

Table of Contents

Part III: Building String Theory: A Theory of Everything............................... 159

Chapter 10: Early Strings and Superstrings: Unearthing the Theory's Beginnings 161

Part IV: The Unseen Cosmos: String Theory On the Boundaries of Knowledge 227

Chapter 13: Making Space for Extra Dimensions................229

Chapter 14: Our Universe — String Theory, Cosmology, and Astrophysics...............245

Introduction

· ·

*W*hy are scientists so excited about string theory? Because string theory is the most likely candidate for a successful theory of quantum gravity — a theory that scientists hope will unite two major physical laws of the universe into one. Right now, these laws (quantum physics and general relativity) describe two totally different types of behavior in totally different ways, and in the realm where neither theory works completely, we really don't know what's going on!

Understanding the implications of string theory means understanding profound aspects of our reality at the most fundamental levels. Are there parallel universes? Is there only one law of nature or infinitely many? Why does our universe follow the laws it does? Is time travel possible? How many dimensions does our universe possess? Physicists are passionately seeking answers to these questions.

Indeed, string theory is a fascinating topic, a scientific revolution that promises to transform our understanding of the universe. As you'll see, these sorts of revolutions have happened before, and this book helps you understand how physics has developed in the past, as well as how it may develop in the future.

This book contains some ideas that will probably, in the coming years, turn out to be completely false. It contains other ideas that may ultimately prove to be fundamental laws of our universe, perhaps forming the foundation for whole new forms of science and technology. No one knows what the future holds for string theory.

About This Book

In this book, I aim to give a clear understanding of the ever-evolving scientific subfield known as string theory. The media is abuzz with talk about this "theory of everything," and when you're done with this book you should know what they're talking about (probably better than they do, most of the time).

In writing this book, I've attempted to serve several masters. First and foremost among them has been scientific accuracy, followed closely by entertainment value. Along the way, I've also done my best to use language that you can understand no matter your scientific background, and I've certainly tried to keep any mathematics to a minimum.

In writing this book, I set out to achieve the following goals:

✔ Provide the information needed to understand string theory (including established physics concepts that predate string theory).

✔ Establish the successes of string theory so far.

✔ Lay out the avenues of study that are attempting to gain more evidence for string theory.

✔ Explore the bizarre (and speculative) implications of string theory.

✔ Present the critical viewpoints in opposition to string theory, as well as some alternatives that may bear fruit if it proves to be false.

✔ Have some fun along the way.

✔ Avoid mathematics at all costs. (You're welcome!)

I hope you, good reader, find that I've been successful at meeting these goals.

And while time may flow in only one direction (Or does it? I explore this in Chapter 16), your reading of this book may not. String theory is a complex scientific topic that has a lot of interconnected concepts, so jumping between concepts is not quite as easy as it may be in some other *For Dummies* reference books. I've tried to help you out by including quick reminders and providing cross-references to other chapters where necessary. So feel free to wander the pages to your heart's content, knowing that if you get lost you can work your way back to the information you need.

Conventions Used in This Book

The following conventions are used throughout the text to make things consistent and easy to understand:

✔ I use monofont for Web sites. **Note:** When this book was printed, some Web addresses may have needed to break across two lines of text. If that happened, rest assured that I haven't put in any extra characters (such as hyphens) to indicate the break. So, when using one of these Web addresses, just type in exactly what you see in this book, as though the line break doesn't exist.

- ✔ I've done my best not to fill the book with technical jargon, which is hard to do in a book on one of the most complex and mathematically driven scientific topics of all time. When I use a technical term, it's in *italics* and closely followed by an easy-to-understand definition.

- ✔ **Bold** is used to highlight key words and phrases in bulleted lists.

Finally, one major convention used in this book is in the title: I use the term "string theory." In Chapter 10, you discover that string theory is actually called *superstring theory*. As you see in Chapter 11, in 1995 physicists realized that the various "string theories" (five existed at the time) included objects other than strings, called *branes*. So, strictly speaking, calling it by the name "string theory" is a bit of a misnomer, but people (including string theorists themselves) do it all the time, so I'm treading on safe ground. Many physicists also use the name *M-theory* to describe string theory after 1995 (although they rarely agree on what the "M" stands for), but, again, I will mostly refer to it just as "string theory" unless the distinction between different types matters.

What You're Not to Read

All the chapters provide you with important information, but some sections offer greater detail or tidbits of information that you can skip for now and come back to later without feeling guilty:

- ✔ **Sidebars:** Sidebars are shaded boxes that give detailed examples or explore a tangent in more detail. Ignoring these won't compromise your understanding of the rest of the material.

- ✔ **Anything with a Technical Stuff icon:** This icon indicates information that's interesting but that you can live without. Read these tidbits later if you're pressed for time.

Foolish Assumptions

About the only assumption that I've made in writing this book is that you're reading it because you want to know something about string theory. I've tried to not even assume that you *enjoy* reading physics books. (I do, but I try not to project my own strangeness on others.)

I have assumed that you have a passing acquaintance with basic physics concepts — maybe you took a physics class in high school or have watched some of the scientific programs about gravity, light waves, black holes, or other physics-related topics on cable channels or your local PBS station. You don't need a degree in physics to follow the explanations in this book, although without a degree in physics you might be amazed that anyone can make sense of any theory this disconnected from our everyday experience. (Even with physics degree, it can boggle the mind.)

As is customary in string theory books for the general public, the mathematics has been avoided. You need a graduate degree in mathematics or physics to follow the mathematical equations at the heart of string theory, and while I have a graduate degree in mathematics, I've assumed that you don't. Don't worry — while a complete understanding of string theory is rooted firmly in the advanced mathematical concepts of quantum field theory, I've used a combination of text and figures to explain the fascinating ideas behind string theory.

How This Book Is Organized

String Theory For Dummies is written so you can easily get to the information you need, read it, and understand it. It's designed to follow the historical development of the theory as much as possible, though many of the concepts in string theory are interconnected. Although I've attempted to make each chapter understandable on its own, I've included cross-references where concepts repeat to get you back to a more thorough discussion of them.

Part I: Introducing String Theory

This first part of the book introduces the key concepts of string theory in a very general way. You read about why scientists are so excited about finding a theory of quantum gravity. Also, you get your first glimpse into the successes and failures of string theory.

Part II: The Physics Upon Which String Theory Is Built

String theory is built upon the major scientific developments of the first 70 years or so of the 20th century. In this part, you find out how physicists (and scientists in general) learn things and what they've learned so far. Part II includes chapters on how science develops, classical physics (before Einstein), Einstein's theory of relativity, quantum physics, and the more recent findings in particle physics and cosmology. The questions raised in these chapters are those that string theory attempts to answer.

Part III: Building String Theory: A Theory of Everything

You get to the heart of the matter in this part. I discuss the creation and development of string theory, from 1968 to early 2009. The amazing transformations of this theory are laid out here. Chapter 12 focuses on ways that the concepts of string theory can be tested.

Part IV: The Unseen Cosmos: String Theory on the Boundaries of Knowledge

Here I take string theory out for a spin in the universe, exploring some of the major concepts in greater detail. Chapter 13 focuses on the concept of extra dimensions, which are at the core of much of string theory study. Chapter 14 explores the implications for cosmology and how string theory could explain certain properties of our universe. Even more amazing, in Chapters 15 and 16, you discover what string theory has to say about possible parallel universes and the potential for time travel.

Part V: What the Other Guys Say: Criticism and Alternatives

The discussion gets heated in this part as you read about the criticisms of string theory. String theory is far from proven, and many scientists feel that it's heading in the wrong direction. Here you find out why and see what alternatives they're posing, such as loop quantum gravity (string theory's biggest competitor). If string theory is wrong, scientists will continue to look for answers to the questions that it seeks to resolve.

Part VI: The Part of Tens

In the *For Dummies* tradition, the final chapters of this book present lists of ten topics. Chapter 20 sums up ten outstanding physics questions that scientists hope any "theory of everything" (including string theory) will answer. Chapter 21 focuses on ten string theorists who have done a lot to advance the field, either through their own research or by introducing string theory concepts to the world through popular books.

Icons Used in this Book

Throughout the book, you'll find icons in the margins that are designed to help you navigate the text. Here's what these icons mean:

Although everything in this book is important, some information is more important than other information. This icon points out information that will definitely be useful later in the book.

In science, theories are often explained with analogies, thought experiments, or other helpful examples that present complex mathematical concepts in a way that is more intuitively understandable. This icon indicates that one of these examples or hints is being offered.

Sometimes I go into detail that you don't need to know to follow the basic discussion and is a bit more technical (or mathematical) than you may be interested in. This icon points out that information, which you can skip without losing the thread of the discussion.

Where to Go from Here

The *For Dummies* books are organized in such a way that you can surf through any of the chapters and find useful information without having to start at Chapter 1. I (naturally) encourage you to read the whole book, but this structure makes it very easy to start with the topics that interest you the most.

If you have no idea what string theory is, then I recommend looking at Chapter 1 as a starting point. If your physics is rusty, pay close attention to Chapters 5–9, which cover the history and current status of the major physics concepts that pop up over and over again.

If you're familiar with string theory but want some more details, jump straight to Chapters 10 and 11, where I explain how string theory came about and reached its current status. Chapter 12 offers some ways of testing the theory, while Chapters 13–16 take concepts from string theory and apply them to some fascinating topics in theoretical physics.

Some of you, however, may want to figure out what all the recent fuss is with people arguing across the blogosphere about string theory. For that, I recommend jumping straight to Chapter 17, which addresses some of the major criticisms of string theory. Chapters 18 and 19 focus heavily on other theories that may either help expand or replace string theory, so they're a good place to go from there.

Part I

Introducing String Theory

The 5th Wave   By Rich Tennant

"Okay—now that the paramedic is here with the defibrillator and smelling salts, prepare to learn about string theory."

In this part . . .

Meet string theory, a bold scientific theory that attempts to reconcile all the physical properties of our universe into a single unified and coherent mathematical framework.

String theory's goal is to make quantum physics and Einstein's theory of gravity (called general relativity) play nice. In this part, I explain why scientists want to find a theory of quantum gravity, and then I review the successes and failures at applying string theory to this search.

This part is something of an overview for the entire book, so stick with me. The foundation laid here may help explain the entire universe.

Chapter 1

So What Is String Theory Anyway?

In This Chapter

▶ Knowing that string theory is based on vibrating strings of energy

▶ Understanding the key elements of string theory

▶ Hoping to explain the entire universe with string theory

▶ Studying string theory could be the driving scientific goal of the 21st century

String theory is a work in progress, so trying to pin down exactly what string theory is, or what the fundamental elements are, can be kind of tricky. Regardless, that's exactly what I try to do in this chapter.

In this chapter, you gain a basic understanding of string theory. I outline the key elements of string theory, which provide the foundation for most of this book. I also discuss the possibility that string theory could be the starting point for a "theory of everything," which would define all of our universe's physical laws in one simple (or not so simple) mathematical formula. Finally, I look at the reasons why you should care about string theory.

String Theory: Seeing What Vibrating Strings Can Tell Us about the Universe

String theory is a physics theory that the universe is composed of vibrating filaments of energy, expressed in precise mathematical language. These *strings* of energy represent the most fundamental aspect of nature. The theory also predicts other fundamental objects, called *branes.* All of the matter in our universe consists of the vibrations of these strings (and branes). One important result of string theory is that gravity is a natural consequence of the theory, which is why scientists believe that string theory may hold the answer to possibly uniting gravity with the other forces that affect matter.

Let me reiterate something important: String theory is a *mathematical* theory. It's based on mathematical equations that can be interpreted in certain ways. If you've never studied physics before, this may seem odd, but *all* physical theories are expressed in the language of mathematics. In this book, I avoid the mathematics and try to get to the heart of what the theory is telling us about the physical universe.

At present, no one knows exactly what the final version of string theory will look like. Scientists have some vague notions about the general elements that will exist within the theory, but no one has come up with the final equation that represents all of string theory in our universe, and experiments haven't yet been able to confirm it (though they haven't successfully refuted it, either). Physicists have created simplified versions of the equation, but it doesn't quite describe our universe . . . yet.

Using tiny and huge concepts to create a theory of everything

String theory is a type of high-energy theoretical physics, practiced largely by particle physicists. It's a *quantum field theory* (see the sidebar "What is quantum field theory?") that describes the particles and forces in our universe based on the way that special extra dimensions within the theory are wrapped up into a very small size (a process called *compactification*). This is the power of string theory — to use the fundamental strings, and the way extra dimensions are compactified, to provide a geometric description of all the particles and forces known to modern physics.

Among the forces needed to be described is, of course, gravity. Because string theory is a quantum field theory, this means that string theory would be a quantum theory of gravity, known as *quantum gravity*. The established theory of gravity, general relativity, has a fluid, dynamic space-time, and one aspect of string theory that's still being worked on is getting this sort of a space-time to emerge out of the theory.

The major achievements of string theory are concepts you can't see, unless you know how to interpret the physics equations. String theory uses no experiments that provide new insights, but it has revealed profound mathematical relationships within the equations, which lead physicists to believe that they must be true. These properties and relationships — called by jargon such as various symmetries and dualities, the cancellation of anomalies, and the explanation of black hole entropy — are described in Chapters 10 and 11.

What is quantum field theory?

Physicists use *fields* to describe the things that don't just have a particular position, but exist at every point in space. For example, you can think about the temperature in a room as a field — it may be different near an open window than near a hot stove, and you could imagine measuring the temperature at every single point in the room. A *field theory,* then, is a set of rules that tell you how some field will behave, such as how the temperature in the room changes over time.

In Chapters 7 and 8, you find out about one of the most important achievements of the 20th century: the development of quantum theory. This refers to principles that lead to seemingly bizarre physical phenomena, which nonetheless seem to occur in the subatomic world.

When you combine these two concepts, you get quantum field theory: a field theory that obeys the principles of quantum theory. All modern particle physics is described by quantum field theories.

In recent years, there has been much public controversy over string theory, waged across headlines and the Internet. These issues are addressed in Part V, but they come down to fundamental questions about how science should be pursued. String theorists believe that their methods are sound, while the critics believe that they are, at best, questionable. Time and experimental evidence will tell which side has made the better argument.

A quick look at where string theory has been

The theory was originally developed in 1968 as a theory that attempted to explain the behavior of *hadrons* (such as protons and neutrons, the particles that make up an atomic nucleus) inside particle accelerators. Physicists later realized this theory could also be used to explain some aspects of gravity.

For more than a decade, string theory was abandoned by most physicists, mainly because it required a large number of extra, unseen dimensions. It rose to prominence again in the mid-1980s, when physicists were able to prove it was a mathematically consistent theory.

In the mid-1990s, string theory was updated to become a more complex theory, called *M-theory,* which contains more objects than just strings. These new objects were called *branes,* and they could have anywhere from zero to nine dimensions. The earlier string theories (which now also include branes) were seen as approximations of the more complete M-theory.

Technically, the modern M-theory is more than the traditional string theory, but the name "string theory" is still often used for M-theory and its various offspring theories. (Even the original superstring theories have been shown to include branes.) My convention in this book is to refer to theories that contain branes, which are variants of M-theory and the original string theories, using the term "string theory."

Introducing the Key Elements of String Theory

Five key ideas are at the heart of string theory and come up again and again. It's best for you to become familiar with these key concepts right off the bat:

- String theory predicts that all objects in our universe are composed of vibrating filaments (and membranes) of energy.

- String theory attempts to reconcile general relativity (gravity) with quantum physics.

- String theory provides a way of unifying all the fundamental forces of the universe.

- String theory predicts a new connection (called *supersymmetry*) between two fundamentally different types of particles, bosons and fermions.

- String theory predicts a number of extra (usually unobservable) dimensions to the universe.

I introduce you to the very basics of these ideas in the following sections.

Strings and branes

When the theory was originally developed in the 1970s, the filaments of energy in string theory were considered to be 1-dimensional objects: strings. (*One-dimensional* indicates that a string has only one dimension, length, as opposed to say a square, which has both length and height dimensions.)

These strings came in two forms — closed strings and open strings. An open string has ends that don't touch each other, while a closed string is a loop with no open end. It was eventually found that these early strings, called Type I strings, could go through five basic types of interactions, as shown in Figure 1-1.

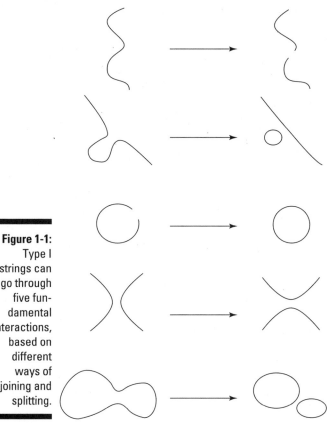

Figure 1-1:
Type I
strings can
go through
five fun-
damental
interactions,
based on
different
ways of
joining and
splitting.

TIP

The interactions are based on a string's ability to have ends join and split apart. Because the ends of open strings can join together to form closed strings, you can't construct a string theory without closed strings.

REMEMBER

This proved to be important, because the closed strings have properties that make physicists believe they might describe gravity! In other words, instead of just being a theory of matter particles, physicists began to realize that string theory may just be able to explain gravity and the behavior of particles.

Over the years, it was discovered that the theory required objects other than just strings. These objects can be seen as sheets, or *branes*. Strings can attach at one or both ends to these branes. A 2-dimensional brane (called a 2-brane) is shown in Figure 1-2. (See Chapter 11 for more about branes.)

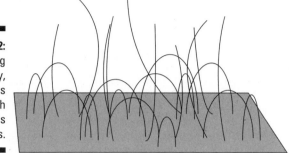

Figure 1-2:
In string theory, strings attach themselves to branes.

Quantum gravity

Modern physics has two basic scientific laws: quantum physics and general relativity. These two scientific laws represent radically different fields of study. *Quantum physics* studies the very smallest objects in nature, while *relativity* tends to study nature on the scale of planets, galaxies, and the universe as a whole. (Obviously, gravity affects small particles too, and relativity accounts for this as well.) Theories that attempt to unify the two theories are theories of *quantum gravity,* and the most promising of all such theories today is string theory.

The closed strings of string theory (see the preceding section) correspond to the behavior expected for gravity. Specifically, they have properties that match the long sought-after *graviton,* a particle that would carry the force of gravity between objects.

Quantum gravity is the subject of Chapter 2, where I cover this idea in much greater depth.

Unification of forces

Hand-in-hand with the question of quantum gravity, string theory attempts to unify the four forces in the universe — electromagnetic force, the strong nuclear force, the weak nuclear force, and gravity — together into one unified theory. In our universe, these fundamental forces appear as four different phenomena, but string theorists believe that in the early universe (when there were incredibly high energy levels) these forces are all described by strings interacting with each other. (If you've never heard of some of these forces, don't worry! They're individually discussed in greater detail in Chapter 2 and throughout Part II.)

Supersymmetry

All particles in the universe can be divided into two types: bosons and fermions. (These types of particles are explained in more detail in Chapter 8.) String theory predicts that a type of connection, called *supersymmetry,* exists between these two particle types. Under supersymmetry, a fermion must exist for every boson and a boson for every fermion. Unfortunately, experiments have not yet detected these extra particles.

Supersymmetry is a specific mathematical relationship between certain elements of physics equations. It was discovered outside of string theory, although its incorporation into string theory transformed the theory into supersymmetric string theory (or superstring theory) in the mid-1970s. (See Chapter 10 for more specifics about supersymmetry.)

One benefit of supersymmetry is that it vastly simplifies string theory's equations by allowing certain terms to cancel out. Without supersymmetry, the equations result in physical inconsistencies, such as infinite values and imaginary energy levels.

Because scientists haven't observed the particles predicted by supersymmetry, this is still a theoretical assumption. Many physicists believe that the reason no one has observed the particles is because it takes a lot of energy to generate them. (Energy is related to mass by Einstein's famous $E = mc^2$ equation, so it takes energy to create a particle.) They may have existed in the early universe, but as the universe cooled off and energy spread out after the big bang, these particles would have collapsed into the lower-energy states that we observe today. (We may not think of our current universe as particularly low energy, but compared to the intense heat of the first few moments after the big bang, it certainly is.)

In other words, the strings vibrating as higher-energy particles lost energy and transformed from one type of particle (one type of vibration) into another, lower-energy type of vibration.

Scientists hope that astronomical observations or experiments with particle accelerators will uncover some of these higher-energy supersymmetric particles, providing support for this prediction of string theory.

Extra dimensions

Another mathematical result of string theory is that the theory only makes sense in a world with more than three space dimensions! (Our universe has three dimensions of space — left/right, up/down, and front/back.) Two possible explanations currently exist for the location of the extra dimensions:

✔ The extra space dimensions (generally six of them) are curled up (*compactified,* in string theory terminology) to incredibly small sizes, so we never perceive them.

✔ We are stuck on a 3-dimensional brane, and the extra dimensions extend off of it and are inaccessible to us.

A major area of research among string theorists is on mathematical models of how these extra dimensions could be related to our own. Some of these recent results have predicted that scientists may soon be able to detect these extra dimensions (if they exist) in upcoming experiments, because they may be larger than previously expected. (See Chapter 13 for more about extra dimensions.)

Understanding the Aim of String Theory

To many, the goal of string theory is to be a "theory of everything" — that is, to be the single physical theory that, at the most fundamental level, describes all of physical reality. If successful, string theory could explain many of the fundamental questions about our universe.

Explaining matter and mass

One of the major goals of current string theory research is to construct a solution of string theory that contains the particles that actually exist in our universe.

String theory started out as a theory to explain particles, such as hadrons, as the different higher vibrational modes of a string. In most current formulations of string theory, the matter observed in our universe comes from the lowest-energy vibrations of strings and branes. (The higher-energy vibrations represent more energetic particles that don't currently exist in our universe.)

The mass of these fundamental particles comes from the ways that these string and branes are wrapped in the extra dimensions that are compactified within the theory, in ways that are rather messy and detailed.

For an example, consider a simplified case where the extra dimensions are curled up in the shape of a donut (called a *torus* by mathematicians and physicists), as in Figure 1-3.

Figure 1-3:
Strings wrap around extra dimensions to create particles with different masses.

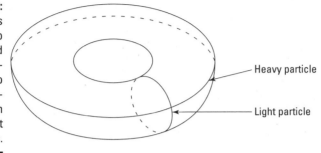

Heavy particle

Light particle

A string has two ways to wrap once around this shape:

- ✔ A short loop around the tube, through the middle of the donut
- ✔ A long loop wrapping around the entire length of the donut (like a string wraps around a yo-yo)

The short loop would be a lighter particle, while the long loop is a heavier particle. As you wrap strings around the torus-shaped compactified dimensions, you get new particles with different masses.

One of the major reasons that string theory has caught on is that this idea — that length translates into mass — is so straightforward and elegant. The compactified dimensions in string theory are much more elaborate than a simple torus, but they work the same way in principle.

It's even possible (though harder to visualize) for a string to wrap in both directions simultaneously — which would, again, give yet another particle with yet another mass. Branes can also wrap around extra dimensions, creating even more possibilities.

Defining space and time

In many versions of string theory, the extra dimensions of space are compactified into a very tiny size, so they're unobservable to our current technology. Trying to look at space smaller than this compactified size would provide results that don't match our understanding of space-time. (As you see in Chapter 2, the behavior of space-time at these small scales is one of the reasons for a search for quantum gravity.) One of string theory's major obstacles is attempting to figure out how space-time can emerge from the theory.

As a rule, though, string theory is built upon Einstein's notion of space-time (see Chapter 6). Einstein's theory has three space dimensions and one time dimension. String theory predicts a few more space dimensions but doesn't change the fundamental rules of the game all that much, at least at low energies.

At present, it's unclear whether string theory can make sense of the fundamental nature of space and time any more than Einstein did. In string theory, it's almost as if the space and time dimensions of the universe are a backdrop to the interactions of strings, with no real meaning on their own.

Some proposals have been developed for how this might be addressed, mainly focusing on space-time as an emergent phenomenon — that is, the space-time comes out of the sum total of all the string interactions in a way that hasn't yet been completely worked out within the theory.

However, these approaches don't meet some physicists' definition, leading to criticism of the theory. String theory's largest competitor, loop quantum gravity, uses the quantization of space and time as the starting point of its own theory, as Chapter 18 explains. Some believe that this will ultimately be another approach to the same basic theory.

Quantizing gravity

The major accomplishment of string theory, if it's successful, will be to show that it's a quantum theory of gravity. The current theory of gravity, general relativity, doesn't allow for the results of quantum physics. Because quantum physics places limitations on the behavior of small objects, it creates major inconsistencies when trying to examine the universe at extremely small scales. (See Chapter 7 for more on quantum physics.)

Unifying forces

Currently, four fundamental forces (more precisely called "interactions" among physicists) are known to physics: gravity, electromagnetic force, weak nuclear force, and strong nuclear force. String theory creates a framework in which all four of these interactions were once a part of the same unified force of the universe.

Under this theory, as the early universe cooled off after the big bang, this unified force began to break apart into the different forces we experience today. Experiments at high energies may someday allow us to detect the unification of these forces, although such experiments are well outside of our current realm of technology.

Appreciating the Theory's Amazing (and Controversial) Implications

Although string theory is fascinating in its own right, what may prove to be even more intriguing are the possibilities that result from it. These topics are explored in greater depth throughout the book and are the focus of Parts III and IV.

Landscape of possible theories

One of the most unexpected and disturbing discoveries of string theory is that instead of one single theory, it turns out there may be a huge number of possible theories (or, more precisely, possible solutions to the theory) — possibly as many as 10^{500} different solutions! (That's a 1 followed by 500 zeroes!) While this huge number has prompted a crisis among some string theorists, others have embraced this as a virtue, claiming that this means that string theory is very rich. In order to wrap their minds around so many possible theories, some string theorists have turned toward the *anthropic principle,* which tries to explain properties of our universe as a result of our presence in it. Still others have no problem with this vast number, actually having expected it and, instead of trying to explain it, just trying to measure the solution that applies to our universe.

With such a large number of theories available, the anthropic principle allows a physicist to use the fact that we're here to choose among only those theories that have physical parameters that allow us to be here. In other words, our very presence dictates the choice of physical law — or is it merely that our presence is an observable piece of data, like the speed of light?

The use of the anthropic principle is one of the most controversial aspects of modern string theory. Even some of the strongest string theory supporters have expressed concern over its application, because of the sordid (and somewhat unscientific) applications to which it has been used in the past and their feeling that all that is needed is an observation of our universe, without anything anthropic applied at all.

As anthropic principle skeptics are quick to point out, physicists only adopt the anthropic principle when they have no other options, and they abandon it if something better comes along. It remains to be seen if string theorists will find another way to maneuver through the string theory landscape. (Chapter 11 has more details about the anthropic principle.)

Parallel universes

Some interpretations of string theory predict that our universe is not the only one. In fact, in the most extreme versions of the theory, an infinite number of other universes exist, some of which contain exact duplicates of our own universe.

As wild as this theory is, it's predicted by current research studying the very nature of the cosmos itself. In fact, parallel universes aren't just predicted by string theory — one view of quantum physics has suggested the theoretical existence of a certain type of parallel universe for more than half a century. In Chapter 15, I explore the scientific concept of parallel universes in greater detail.

Wormholes

Einstein's theory of relativity predicts warped space called a *wormhole* (also called an *Einstein-Rosen bridge*). In this case, two distant regions of space are connected by a shorter wormhole, which gives a shortcut between those two distant regions, as shown in Figure 1-4.

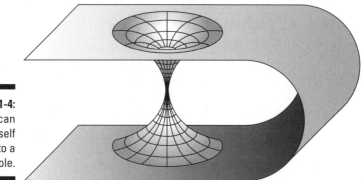

Figure 1-4:
Space can warp itself into a wormhole.

String theory allows for the possibility that wormholes extend not only between distant regions of our own universe, but also between distant regions of parallel universes. Perhaps universes that have different physical laws could even be connected by wormholes. (Chapters 15 and 16 contain more info on wormholes.)

In fact, it's not clear whether wormholes will exist within string theory at all. As a quantum gravity theory, it's possible that the general relativity solutions that give rise to potential wormholes might go away.

The universe as a hologram

In the mid-1990s, two physicists came up with an idea called the *holographic principle*. In this theory, if you have a volume of space, you can take all the information contained in that space and show that it corresponds to information "written" on the surface of the space. As odd as it seems, this holographic principle may be key in resolving a major mystery of black holes that has existed for more than 20 years!

Many physicists believe that the holographic principle will be one of the fundamental physical principles that will allow insights into a greater understanding of string theory. (Check out Chapter 11 for more on the holographic principle.)

Time travel

Some physicists believe that string theory may allow for multiple dimensions of time (by no means the dominant view). As our understanding of time grows with string theory, it's possible that scientists may discover new means of traveling through the time dimension or show that such theoretical possibilities are, in fact, impossible, as most physicists believe. (Flip to Chapter 16 if you're ready to make your time travel reservation.)

The big bang

String theory is being applied to cosmology, which means that it may give us insights into the formation of the universe. The exact implications are still being explored, but some believe that string theory supports the current cosmological model of inflation, while others believe it allows for entirely universal creation scenarios.

Inflation theory predicts that, very shortly after the original big bang, the universe began to undergo a period of rapid, exponential inflation. This theory, which applies principles of particle physics to the early universe as a whole, is seen by many as the only way to explain some properties of the early universe.

In string theory, there also exists a possible alternate model to our current big bang model in which two branes collided together and our universe is the result. In this model, called the *ekpyrotic universe,* the universe goes through cycles of creation and destruction, over and over. (Chapter 14 covers the big bang theory and the ekpyrotic universe.)

The end of the universe

The ultimate fate of the universe is a question that physics has long explored, and a final version of string theory may help us ultimately determine the matter density and cosmological constant of the universe. By determining these values, cosmologists will be able to determine whether our universe will ultimately contract in upon itself, ending in a big crunch — and perhaps start all over again. (See Chapter 14 for more on these speculations.)

Why Is String Theory So Important?

String theory yields many fascinating subjects for thought, but you may be wondering about the practical importance of it. For one thing, string theory is the next step in our growing understanding of the universe. If that's not practical enough, then there's this consideration: Your tax money goes to fund scientific research, and the people trying to get that money want to use it to study string theory (or its alternatives).

A completely honest string theorist would be forced to say that there are probably no practical applications for string theory, at least in the foreseeable future. This doesn't look that great on either the cover of a book or a magazine column, so it gets spiced up with talk about parallel universes, extra time dimensions, and discovering new fundamental symmetries of nature. They might exist, but the theory's predictions make it so that they're unlikely to ever be particularly useful, so far as we know.

Understanding the nature of the universe better is a good goal in its own right — as old as humanity, some might say — but when you're looking at funding multibillion dollar particle accelerators or research satellite programs, you might want something tangible for your money and, unfortunately, there's no reason to think that string theory is going to give you anything practical.

Does this mean that exploring string theory isn't important? No, and it's my hope that reading Part II of this book will help illuminate the key at the heart of the search for string theory, or any new scientific truth.

No one knows where a scientific theory will lead until the theory is developed and tested.

In 1905, when Albert Einstein first presented his famous equation $E = mc^2$, he thought it was an intriguing relationship but had no idea that it would result in something as potent as the atomic bomb. He had no way of knowing the corrections to time calculations demanded by special relativity and general relativity would someday be required to get the worldwide global positioning system (GPS) to operate correctly (as discussed in Chapter 6).

Quantum physics, which on the surface is about as theoretical of a study as they come, is the basis for the laser and transistor, two pieces of technology that are at the heart of modern computers and communication systems.

Even though we don't know what a purely theoretical concept like string theory may lead to, history has shown that it will almost certainly lead somewhere profound.

For an example of the unexpected nature of scientific progress, consider the discovery and study of electricity, which was originally seen as a mere parlor trick. You could predict some technologies from the discovery of electricity, to be sure, such as the light bulb. But some of the most profound discoveries are things that may never have been predicted — radio and television, the computer, the Internet, the cellphone, and so on.

The impact of science extends into culture as well. Another byproduct of electricity is rock and roll music, which was created with the advent of electric guitars and other electric musical instruments.

If electricity can lead to rock and roll and the Internet, then imagine what sort of unpredicted (and potentially unpredictable) cultural and technological advances string theory could lead to!

Chapter 2

The Physics Road Dead Ends at Quantum Gravity

*P*hysicists like to group concepts together into neat little boxes with labels, but sometimes the theories they try to put together just don't want to get along. Right now, nature's fundamental physical laws can fit into one of two boxes: general relativity or quantum physics. But concepts from one box just don't work together well with concepts from the other box.

Any theory that can get these two physics concepts to work together would be called a *theory of quantum gravity*. String theory is currently the most likely candidate for a successful theory of quantum gravity.

In this chapter, I explain why scientists want (and need) a theory of quantum gravity. I begin by giving an overview of the scientific understanding of gravity, which is defined by Einstein's theory of general relativity, and our understanding of matter and the other forces of nature, in terms of quantum mechanics. With these fundamental tools in place, I then explain the ways in which these two theories clash with each other that provides the basis for quantum gravity. Finally, I outline various attempts to unify these theories and the forces of physics together into one coherent system, and the failures they've run into.

Understanding Two Schools of Thought on Gravity

Physicists are searching for a theory of quantum gravity because the current laws governing gravity don't work in all situations. Specifically, the theory of gravity seems to "break down" (that is, the equations become physically meaningless) in certain circumstances that I describe later in the chapter. To understand what this means, you must first understand a bit about what physicists know about gravity.

Gravity is an attractive force that binds objects together, seemingly across any amount of distance. The formulation of the classical theory of gravity by Sir Isaac Newton was one of the greatest achievements of physics. Two centuries later, the reinvention of gravity by Albert Einstein placed him in the pantheon of indisputably great scientific thinkers of all time.

Unless you're a physicist, you probably take gravity for granted. It's an amazing force, able to hold the heavens together while being overcome by my 3-year-old when he's on a swing — but not for long. At the scale of an atom, gravity is irrelevant compared to the electromagnetic force. In fact, a simple magnet can overcome the entire force of the planet Earth to pick up metallic objects, from paper clips to automobiles.

Newton's law of gravity: Gravity as force

Sir Isaac Newton developed his theory of gravity in the late 1600s. This amazing theory involved bringing together an understanding of astronomy and the principles of motion (known as *mechanics* or *kinematics*) into one comprehensive framework that also required the invention of a new form of mathematics: calculus. In Newton's gravitational theory, objects are drawn together by a physical force that spans vast distances of space.

The key is that gravity binds all objects together (much like the Force in *Star Wars*). The apple falling from a tree and the moon's motion around Earth are two manifestations of the exact same fundamental force.

The relationship that Newton discovered was a mathematical relationship (he did, after all, have to invent calculus to get it all to work out), just like relativity, quantum mechanics, and string theory.

In Newton's gravitational theory, the force between two objects is based on the product of their masses, divided by the square of the distance between them. In other words, the heavier the two objects are, the more force there is between them, assuming the distance between them stays the same. (See the nearby sidebar "A matter of mass" for clarification of this relationship.)

A matter of mass

When I say that the force between objects is proportional to the mass of the two objects, you may think this means that heavier things fall faster than lighter things. For example, wouldn't a bowling ball fall faster than a soccer ball?

In fact, as Galileo showed (though not with modern bowling and soccer balls) years before Newton was born, this isn't the case. For centuries, most people had assumed that heavier objects fell faster than light objects. Newton was aware of Galileo's results, which was why he was able to figure out how to define force the way he did.

By Newton's explanation, it takes more force to move a heavier object. If you dropped a bowling ball and soccer ball off a building (which I don't recommend), they would accelerate at the exact same rate (ignoring air resistance) — approximately 9.8 meters per second.

The force acting between the bowling ball and Earth would be higher than the force acting on the soccer ball, but because it takes more force to get the bowling ball moving, the actual rate of acceleration between the two is identical.

Realistically, if you performed the experiment there would be a slight difference. Because of air resistance, the lighter soccer ball would probably be slowed down if dropped from a high enough point, while the bowling ball would not. But a properly constructed experiment, in which air resistance is completely neutralized (such as in a vacuum), shows that the objects fall at the same rate, regardless of mass.

The fact that the force is divided by distance squared means that if the same two objects are closer to each other, the power of gravity increases. If the distance gets wider, the force drops. The inverse square relationship means that if the distance doubles, the force drops to one-fourth of its original intensity. If the distance is halved, the force increases by four times.

If the objects are very far away, the effect of gravity becomes very small. The reason gravity has any impact on the universe is because there's *a lot* of it. Gravity itself is very weak, as forces go.

The opposite is true, as well, and if two objects get extremely close to each other — and I'm talking *extremely* close here — then gravity can become incredibly powerful, even among objects that don't have much mass, like the fundamental particles of physics.

This isn't the only reason gravity is observed so much. Gravity's strength in the universe also comes from the fact that it's always attracting objects together. The electromagnetic force sometimes attracts objects and sometimes repulses them, so on the scale of the universe at large, it tends to counteract itself. Finally, gravity interacts at very large distances, as opposed to some other forces (the nuclear forces) that only work at distances smaller than an atom.

I delve a bit deeper into Newton's work, both in gravity and in other related areas, in Chapter 5.

Despite the success of Newton's theory, he had a few nagging problems in the back of his mind. First and foremost among those was the fact that though he had a model for gravity, he didn't know *why* gravity worked. The gravity that he described was an almost mystical force (like the Force!), acting across great distances with no real physical connection required. It would take two centuries and Albert Einstein to resolve this problem.

Einstein's law of gravity: Gravity as geometry

Albert Einstein would revolutionize the way physicists saw gravity. Instead of gravity as a force acting between objects, Einstein instead envisioned a universe in which each object's mass caused a slight bending of space (actually space-time) around it. The movement of an object along the shortest distance in this space-time was gravity. Instead of being a force, gravity was actually an effect of the geometry of space-time itself.

Einstein proposed that motion in the universe could be explained in terms of a coordinate system with three space dimensions — up/down, left/right, and backward/forward, for example — and one time dimension. This 4-dimensional coordinate system, developed by Einstein's old professor Hermann Minkowski, was called *space-time,* and came out of Einstein's 1905 *theory of special relativity*.

As Einstein generalized this theory, creating the *theory of general relativity* in 1916, he was able to include gravity in his explanations of motion. In fact, the concept of space-time was crucial to it. The space-time coordinate system bent when matter was placed in it. As objects moved within space and time, they naturally tried to take the shortest path through the bent space-time.

We follow our orbit around the sun because it's the shortest path (called a *geodesic* in mathematics) through the curved space-time around the sun.

Einstein's relativity is covered in depth in Chapter 6, and the major implications of relativity to the evolution of the universe are covered in Chapter 9. The space-time dimensions are discussed in Chapter 13.

Describing Matter: Physical and Energy-Filled

Einstein helped to revolutionize our ideas about the composition of matter as much as he did about space, time, and gravity. Thanks to Einstein, scientists realize that mass — and therefore matter itself — is a form of energy. This realization is at the heart of modern physics. Because gravity is an interaction between objects made up of matter, understanding matter is crucial to understanding why physicists need a theory of quantum gravity.

Viewing matter classically: Chunks of stuff

The study of matter is one of the oldest physics disciplines, because philosophers tried to understand what made up objects. Even fairly recently, a physical understanding of matter was elusive, as physicists debated the existence of *atoms* — tiny, indivisible chunks of matter that couldn't be broken up anymore.

One key physics principle was that matter could be neither created nor destroyed, but could only change from one form to another. This principle is known as the *conservation of mass*.

Though it can't be created or destroyed, matter can be broken, which led to the question of whether there was a smallest chunk of matter, the atom, as the ancient Greeks had proposed — a question that, throughout the 1800s, seemed to point toward an affirmative answer.

As an understanding of *thermodynamics* — the study of heat and energy, which made things like the steam engine (and the Industrial Revolution) possible — grew, physicists began to realize that heat could be explained as the motion of tiny particles.

The atom had returned, though the findings of 20th-century quantum physics would reveal that the atom wasn't indivisible as everyone thought.

Viewing matter at a quantum scale: Chunks of energy

With the rise of modern physics in the 20th century, two key facts about matter became clear:

- ✔ As Einstein had proposed with his famous $E = mc^2$ equation, matter and energy are, in a sense, interchangeable.
- ✔ Matter was incredibly complex, made up of an array of bizarre and unexpected types of particles that joined together to form other types of particles.

The atom, it turned out, was composed of a nucleus surrounded by electrons. The nucleus was made up of protons and neutrons, which were, in turn, made up of strange new particles called quarks! As soon as physicists thought they had reached a fundamental unit of matter, they seemed to discover that it could be broken open and still smaller units could be pulled out.

Not only that, but even these fundamental particles didn't seem to be enough. It turned out that there were three families of particles, some of which only appeared at significantly higher energies than scientists had previously explored.

Today, the Standard Model of particle physics contains 18 distinct fundamental particles, 17 of which have been observed experimentally. (Physicists are still waiting on the Higgs boson.)

Grasping for the Fundamental Forces of Physics

Even while the numbers of particles became more bizarre and complex, the ways those objects interacted turned out to be surprisingly straightforward. In the 20th century, scientists discovered that objects in the universe experienced only four fundamental types of interactions:

- Electromagnetism
- Strong nuclear force
- Weak nuclear force
- Gravity

Physicists have discovered profound connections between these forces — except for gravity, which seems to stand apart from the others for reasons that physicists still aren't completely certain about. Trying to incorporate gravity with all the other forces — to discover how the fundamental forces are related to each other — is a key insight that many physicists hope a theory of quantum gravity will offer.

Electromagnetism: Super-speedy energy waves

Discovered in the 19th century, the *electromagnetic force* (or *electromagnetism*) is a unification of the electrostatic force and the magnetic force. In the mid-20th century, this force was explained in a framework of quantum mechanics called *quantum electrodynamics,* or *QED.* In this framework, the electromagnetic force is transferred by particles of light, called *photons.*

The relationship between electricity and magnetism is covered in Chapter 5, but the basic relationship comes down to electrical charge and its motion. The electrostatic force causes charges to exert forces on each other in a relationship that's similar to (but more powerful than) gravity — an inverse square law. This time, though, the intensity is based not on the mass of the objects, but the charge.

The *electron* is a particle that contains a negative electrical charge, while the *proton* in the atomic nucleus has a positive electrical charge. Traditionally, electricity is seen as the flow of electrons (negative charge) through a wire. This flow of electrons is called an *electric current.*

A wire with an electrical current flowing through it creates a magnetic field. Alternately, when a magnet is moved near a wire, it causes a current to flow. (This is the basis of most electric power generators.)

This is the way in which electricity and magnetism are related. In the 1800s, physicist James Clerk Maxwell unified the two concepts into one theory, called *electromagnetism,* which depicted this force as waves of energy moving through space.

One key component of Maxwell's unification was a discovery that the electromagnetic force moved at the speed of light. In other words, the electromagnetic waves that Maxwell predicted from his theory were a form of light waves.

Quantum electrodynamics retains this relationship between electromagnetism and light, because in QED the information about the force is transferred between two charged particles (or magnetic particles) by another particle — a *photon,* or particle of light. (Physicists say that the electromagnetic force is *mediated* by a photon.)

Nuclear forces: What the strong force joins, the weak force tears apart

In addition to gravity and electromagnetism, 20th-century physics discovered two nuclear forces called the *strong nuclear force* and *weak nuclear force.* These forces are also mediated by particles. The strong force is mediated by a type of particle called a *gluon.* The weak force is mediated by three particles: Z, W^+, and W^- *bosons.* (You can read more about these particles in Chapter 8.)

The strong nuclear force holds quarks together to form protons and neutrons, but it also holds the protons and neutrons together inside the atom's nucleus.

The weak nuclear force, on the other hand, is responsible for radioactive decay, such as when the neutron decays into a proton. The processes governed by the weak nuclear force are responsible for the burning of stars and the formation of heavy elements inside of stars.

Infinities: Why Einstein and the Quanta Don't Get Along

Einstein's theory of general relativity, which explains gravity, does an excellent job at explaining the universe on the scale of the cosmos. Quantum physics does an excellent job of explaining the universe on the scale of an atom or smaller. In between those scales, good old-fashioned classical physics usually rules.

Unfortunately, some problems bring general relativity and quantum physics into conflict, resulting in mathematical infinities in the equations. (Infinity is essentially an abstract number that is larger than any other numbers. Though certain cartoon characters like to go "To infinity and beyond," scientists don't like to see infinities come up in mathematical equations.) Infinities come up in quantum physics, but physicists have developed mathematical techniques to tame them in many of those cases, so the results match experiments. In some cases, however, these techniques don't apply. Because physicists never witness real infinities in nature, these troublesome problems motivate a search for quantum gravity.

Each of the theories works fine on its own, but when you get into areas where both have something specific to say about the same thing — such as what's going on at the border of a black hole — things get very complicated. The quantum fluctuations make the distinction between the inside and outside of the black hole kind of fuzzy, and general relativity needs that distinction to work properly. Neither theory by itself can fully explain what's going on in these specific cases.

This is the heart of why physicists need a theory of quantum gravity. With the current theories, you get situations that don't look like they make sense. Physicists don't see infinities, yet as you'll see, both relativity and quantum physics indicate that they should exist. Reconciling this bizarre region in the middle, where neither theory can fully describe what's going on, is the goal of quantum gravity.

Singularities: Bending gravity to the breaking point

Because matter causes a bending of space-time, cramming a lot of matter into a very small space causes a lot of bending of space-time. In fact, some solutions to Einstein's general relativity equations show situations where space-time bends an infinite amount — called a *singularity*. Specifically, a space-time singularity shows up in the mathematical equations of general relativity in two situations:

- ✔ During the early big bang period of the universe's history
- ✔ Inside black holes

These subjects are covered in more detail in Chapter 9, but both situations involve a density of matter (a lot of matter in a small space) that's enough to cause problems with the smooth space-time geometry that relativity depends on.

These singularities represent points where the theory of general relativity breaks down completely. Even talking about what goes on at this point becomes meaningless, so physicists need to refine the theory of gravity to include rules about how to talk about these situations in a meaningful way.

Some believe that this problem can be solved by altering Einstein's theory of gravity (as you see in Chapter 19). String theorists don't usually want to modify gravity (at least at the energy levels scientists normally look at); they just want to create a framework that allows gravity to work without running into these mathematical (and physical) infinities.

Quantum jitters: Space-time under a quantum microscope

A second type of infinity, proposed by John Wheeler in 1955, is the *quantum foam* or, as it's called by string theorist and best-selling author Brian Greene, the *quantum jitters*. Quantum effects mean that space-time at very tiny distance scales (called the *Planck length*) is a chaotic sea of virtual particles being created and destroyed. At these levels, space-time is certainly not smooth as relativity suggests, but is a tangled web of extreme and random energy fluctuations, as shown in Figure 2-1.

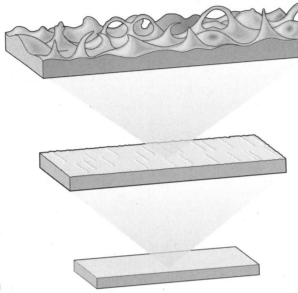

Figure 2-1:
If you
zoom in on
space-time
enough,
a chaotic
"quantum
foam"
may exist.

The basis for the quantum jitters is the *uncertainty principle,* one of the key (and most unusual) features of quantum physics. This is explained in more detail in Chapter 7, but the key component of the uncertainty principle is that certain pairs of quantities — for example, position and velocity, or time and energy — are linked together, so that the more precisely one is measured, the more uncertain the other quantity is. This isn't just a statement about measurement, though, but a fundamental uncertainty in nature!

In other words, nature is a bit "blurry" according to quantum physics. This blurriness only shows up at very small distances, but this problem creates the quantum foam.

One example of the blurriness comes in the form of virtual particles. According to quantum field theory (a *field theory* is one where each point in space has a certain value, similar to a gravitational field or electromagnetic field), even the empty void of space has a slight energy associated with it. This energy can be used to, very briefly, bring a pair of particles — a particle and its antiparticle, to be precise — into existence. The particles exist for only a moment, and then destroy each other. It's as if they borrowed enough energy from the universe to exist for just a few fractions of a second.

The problem is that when you look at space-time at very small scales, the effects of these virtual particles become very important. The energy fluctuations predicted by the uncertainty principle take on massive proportions. Without a quantum theory of gravity, there's no way to really figure out what's going on at sizes that small.

Unifying the Forces

The attempt to unite gravity with the other three forces, as well as with quantum physics, was one of the driving forces of physics throughout the 20th century (and it still is). In a way, these sorts of unifications of different ideas are the major discoveries in science throughout the ages.

Quantum electrodynamics successfully created a quantum theory of electromagnetism. Later, the electroweak theory unified this theory together with the weak nuclear force. The strong nuclear force is explained by quantum chromodynamics. The current model of physics that explains all three of these forces is called the *Standard Model of particle physics,* which is covered in much more detail in Chapter 8. Unifying gravity with the other forces would create a new version of the Standard Model and would explain how gravity works on the quantum level. Many physicists hope that string theory will ultimately prove to be this theory.

Einstein's failed quest to explain everything

After Einstein successfully worked the major kinks out of his theory of general relativity, he turned his attention toward trying to unify this theory of gravity with electromagnetism, as well as with quantum physics. In fact, Einstein would spend most of the rest of his life trying to develop this unified theory, but would die unsuccessful.

Throughout the quest, Einstein looked at almost any theory he could think of. One of these ideas was to add an extra space dimension and roll it up into a very small size. This approach, called a Kaluza-Klein theory after the men who created it, is addressed in Chapter 6. This same approach would eventually be used by string theorists to deal with the pesky extra dimensions that arose in their own theories.

Ultimately, none of Einstein's attempts bore fruit. To the day of his death, he worked feverishly on completing his unified field theory in a manner that many physicists have considered a sad end to such a great career.

Today, however, some of the most intense theoretical physics work is in the search for a theory to unify gravity and the rest of physics, mainly in the form of string theory.

A particle of gravity: The graviton

The Standard Model of particle physics explains electromagnetism, the strong nuclear force, and the weak nuclear force as fields that follow the rules of gauge theory. *Gauge theory* is based heavily on mathematical symmetries. Because these forces are quantum theories, the gauge fields come in discrete units (that's where the word quantum comes from) — and these units actually turn out to be particles in their own right, called *gauge bosons.* The forces described by a gauge theory are carried, or *mediated,* by these gauge bosons. For example, the electromagnetic force is mediated by the proton. When gravity is written in the form of a gauge theory, the gauge boson for gravity is called the *graviton.* (If you're confused about gauge theories, don't worry too much — just remember that the graviton is what makes gravity work and you'll know everything that you need to know to understand their application to string theory.)

Physicists have identified some features of the theoretical graviton so that, if it exists, it can be recognized. For one thing, the particle is *massless,* which means it has no rest mass — the particle is always in motion, and that probably means it travels at the speed of light (although in Chapter 19 you find out about a theory of modified gravity in which gravity and light move through space at different speeds).

Another feature of the graviton is that it has a spin of 2. (*Spin* is a quantum number indicating an inherent property of a particle that acts kind of like angular momentum. Fundamental particles have an inherent spin, meaning that they interact with other particles like they're spinning even when they aren't.)

A graviton also has no electrical charge. It's a stable particle, which means it would not decay.

So physicists are looking for a massless particle moving at an incredibly fast speed, with no electrical charge, and a quantum spin of 2. Even though the graviton has never been discovered by experiment, it's the gauge boson that mediates the gravitational force. Given the incredibly weak strength of gravity in relation to other forces, trying to identify gravitons is an incredibly hard task.

The possible existence of the graviton in string theory is one of the major motivations for looking toward the theory as a likely solution to the problem of quantum gravity.

Supersymmetry's role in quantum gravity

Supersymmetry is a principle that says that two types of fundamental particles, bosons and fermions, are connected to each other. The benefit of this type of symmetry is that the mathematical relationships in gauge theory reduce in such a way that unifying all the forces becomes more feasible. (I explain bosons and fermions in greater detail in Chapter 8, while I present a more detailed discussion of supersymmetry in Chapter 10.)

The top graph in Figure 2-2 shows the three forces described by the Standard Model modeled at different energy levels. If the three forces met up in the same point, it would indicate that there might be an energy level where these three forces became fully unified into one superforce.

However, as seen in the lower graph of Figure 2-2, when supersymmetry is introduced into the equation (literally, not just metaphorically), the three forces meet in a single point. If supersymmetry proves to be true, it's strong evidence that the three forces of the Standard Model unify at high enough energy.

Many physicists believe that all four forces were once unified at high energy levels, but as the universe reduced into a lower-energy state, the inherent symmetry between the forces began to break down. This broken symmetry caused the creation of four distinct forces of nature.

The goal of a theory of quantum gravity is, in a sense, an attempt to look back in time, to when these four forces were unified as one. If successful, it would profoundly affect our understanding of the first few moments of the universe — the last time that the forces joined together in this way.

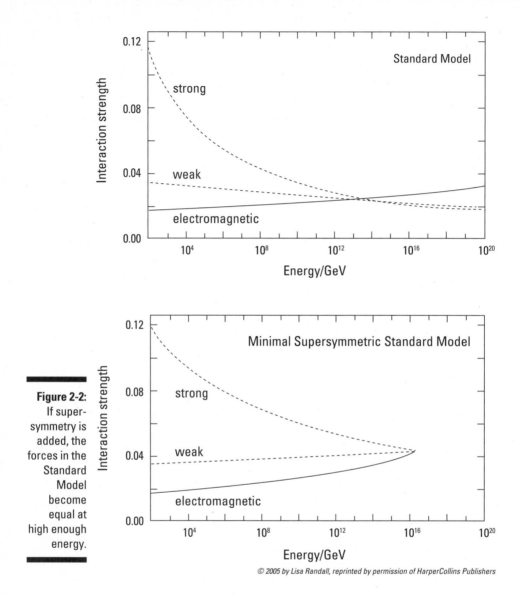

Figure 2-2:
If super-symmetry is added, the forces in the Standard Model become equal at high enough energy.

Chapter 3

Accomplishments and Failures of String Theory

. . . .

In This Chapter

▶ Embracing string theory's achievements

▶ Poking holes in string theory

▶ Wondering what the future of string theory holds

. . . .

String theory is a work in progress, having captured the hearts and minds of much of the theoretical physics community while being apparently disconnected from any realistic chance of definitive experimental proof. Despite this, it has had some successes — unexpected predictions and achievements that may well indicate string theorists are on the right track.

String theory critics would also point out (and many string theorists would probably agree) that the last decade hasn't been kind to string theory because the momentum toward a unified theory of everything has slowed because of a fracture among many different versions of string theory, instead of a single version of the theory.

In this chapter, you see some of the major successes and failures of string theory, as well as look at the possibilities for where string theory may go from here. The controversy over string theory rests entirely on how much significance physicists give to these different outcomes.

Celebrating String Theory's Successes

String theory has gone through many transformations since its origins in 1968 when it was hoped to be a model of certain types of particle collisions. It initially failed at that goal, but in the 40 years since, string theory has developed into the primary candidate for a theory of quantum gravity. It has driven major developments in mathematics, and theorists have used insights from string theory to tackle other, unexpected problems in physics. In fact, the very presence of gravity within string theory is an unexpected outcome!

Predicting gravity out of strings

The first and foremost success of string theory is the unexpected discovery of objects within the theory that match the properties of the graviton. These objects are a specific type of closed strings that are also massless particles that have spin of 2, exactly like gravitons. To put it another way, gravitons are a spin-2 massless particle that, under string theory, can be formed by a certain type of vibrating closed string. String theory wasn't created to have gravitons — they're a natural and required consequence of the theory.

One of the greatest problems in modern theoretical physics is that gravity seems to be disconnected from all the other forces of physics that are explained by the Standard Model of particle physics. String theory solves this problem because it not only includes gravity, but it makes gravity a necessary byproduct of the theory.

Explaining what happens to a black hole (sort of)

A major motivating factor for the search for a theory of quantum gravity is to explain the behavior of black holes, and string theory appears to be one of the best methods of achieving that goal. String theorists have created mathematical models of black holes that appear similar to predictions made by Stephen Hawking more than 30 years ago and may be at the heart of resolving a long-standing puzzle within theoretical physics: What happens to matter that falls into a black hole?

Scientists' understanding of black holes has always run into problems, because to study the quantum behavior of a black hole you need to somehow describe all the *quantum states* (possible configurations, as defined by quantum physics) of the black hole. Unfortunately, black holes are objects in general relativity, so it's not clear how to define these quantum states. (See Chapter 2 for an explanation of the conflicts between general relativity and quantum physics.)

String theorists have created models that appear to be identical to black holes in certain simplified conditions, and they use that information to calculate the quantum states of the black holes. Their results have been shown to match Hawking's predictions, which he made without any precise way to count the quantum states of the black hole.

This is the closest that string theory has come to an experimental prediction. Unfortunately, there's nothing experimental about it because scientists can't directly observe black holes (yet). It's a theoretical prediction that unexpectedly matches another (well-accepted) theoretical prediction about black holes. And, beyond that, the prediction only holds for certain types of black holes and has not yet been successfully extended to all black holes.

For a more extended look at black holes and string theory, check out Chapters 9, 11, and 14.

Explaining quantum field theory using string theory

One of the major successes of string theory is something called the Maldacena conjecture, or the AdS/CFT correspondence. (I get into what this stands for and means in Chapter 11.) Developed in 1997 and soon expanded on, this correspondence appears to give insights into gauge theories, such as those at the heart of quantum field theory. (See Chapter 2 for an explanation of gauge theories.)

The original AdS/CFT correspondence, written by Juan Maldacena, proposes that a certain 3-dimensional (three space dimensions, like our universe) gauge theory, with the most supersymmetry allowed, describes the same physics as a string theory in a 4-dimensional (four space dimensions) world. This means that questions about string theory can be asked in the language of gauge theory, which is a quantum theory that physicists know how to work with!

Like John Travolta, string theory keeps making a comeback

String theory has suffered more setbacks than probably any other scientific theory in the history of the world, but those hiccups don't seem to last that long. Every time it seems that some flaw comes along in the theory, the mathematical resiliency of string theory seems to not only save it, but to bring it back stronger than ever.

When extra dimensions came into the theory in the 1970s, the theory was abandoned by many, but it had a comeback in the first superstring revolution. It then turned out there were five distinct versions of string theory, but a second superstring revolution was sparked by unifying them. When string theorists realized a vast number of solutions of string theories (each solution to string theory is called a *vacuum,* while many solutions are called *vacua*) were possible, they turned this into a virtue instead of a drawback. Unfortunately, even today, some scientists believe that string theory is failing at its goals. (See "Considering String Theory's Setbacks" later in this chapter.)

Being the most popular theory in town

Many young physicists feel that string theory, as the primary theory of quantum gravity, is the best (or only) avenue for making a significant contribution to our understanding of this topic. Over the last two decades, high-energy theoretical physics (especially in the United States) has become dominated by string theorists. In the high-stakes world of "publish or perish" academia, this is a major success.

Why do so many physicists turn toward this field when it offers no experimental evidence? Some of the brightest theoretical physicists of either the 20th or the 21st centuries — Edward Witten, John Henry Schwarz, Leonard Susskind, and others you meet throughout this book — continually return to the same common reasons in support of their interest:

- ✔ If string theory were wrong, it wouldn't provide the rich structure that it does, such as with the development of the heterotic string (see Chapter 10) that allows for an approximation of the Standard Model of physics within string theory.

- ✔ If string theory were wrong, it wouldn't lead to better understandings of quantum field theory, quantum chromodynamics (see Chapter 8), or the quantum states of black holes, as have been presented by the work of Leonard Susskind, Andrew Strominger, Cumrun Vafa, and Juan Maldacena (see Chapters 11 and 14).

- ✔ If string theory were wrong, it would have collapsed in upon itself well before now, instead of passing many mathematical consistency checks (such as those discussed in Chapter 10) and providing more and more elaborate ways to be interpreted, such as the dualities and symmetries that allowed for the presentation of M-theory (as discussed in Chapter 11).

This is how theoretical physicists think, and it's why so many of them continue to believe that string theory is the place to be. The mathematical beauty of the theory, the fact that it's so adaptable, is seen as one of its virtues. The theory continues to be refined, and it hasn't been shown to be incompatible with our universe. There has been no brick wall where the theory failed to provide something new and (in some eyes, at least) meaningful, so those studying string theory have had no reason to give up and look somewhere else. (The history of string theory in Chapters 10 and 11 offers a better appreciation of these achievements.)

Whether this resilience of string theory will translate someday into proof that the theory is fundamentally correct remains to be seen, but for the majority of those working on the problems, confidence is high.

As you can read in Chapter 17, this popularity is also seen by some critics as a flaw. Physics thrives on the rigorous debate of conflicting ideas, and some physicists are concerned that the driving support of string theory, to the exclusion of all other ideas, isn't healthy for the field. For some of these critics, the mathematics of string theory has, indeed, already shown that the theory isn't performing as expected (or, in their view, as needed to be a fundamental theory) and the string theorists are in denial.

Considering String Theory's Setbacks

Because string theory has made so few specific predictions, it's hard to disprove it, but the theory has fallen short of some of the hype about how it will be a fundamental theory to explain all the physics in our universe, a "theory of everything." This failure to meet that lofty goal seems to be the basis of many (if not most) of the attacks against it.

In Chapter 17, you find more detailed criticisms of string theory. Some of these cut to the very heart of whether string theory is even scientific or whether it's being pursued in the correct way. For now, I leave these more abstract questions and focus on three issues that even most string theorists aren't particularly happy about:

✔ Because of supersymmetry, string theory requires a large number of particles beyond what scientists have ever observed.

✔ This new theory of gravity was unable to predict the accelerated expansion of the universe that was detected by astronomers.

✔ A vastly large number of mathematically feasible string theory *vacua* (solutions) currently exist, so it seems virtually impossible to figure out which could describe our universe.

The following sections cover these dilemmas in more detail.

The universe doesn't have enough particles

For the mathematics of string theory to work, physicists have to assume a symmetry in nature called *supersymmetry,* which creates a correspondence between different types of particles. One problem with this is that instead of the 18 fundamental particles in the Standard Model, supersymmetry requires at least 36 fundamental particles (which means that nature allows 18 particles that scientists have never seen!).

In some ways, string theory does make things simpler — the fundamental objects are *strings* and *branes* or, as predicted by matrix theory, zero-dimensional branes called *partons*. These strings, branes, or possibly partons make up the particles that physicists have observed (or the ones they hope to observe). But that's on a very fundamental level; from a practical standpoint, string theory doubles the number of particles allowed by nature from 18 to 36.

One of the biggest possible successes for string theory would be to experimentally detect these missing supersymmetric partner particles. The hope of many theoretical physicists is that when the Large Hadron Collider particle accelerator at CERN in Switzerland goes fully online, it will detect supersymmetric particles.

Even if successful, proof of supersymmetry doesn't inherently prove string theory, so the debate would continue to rage on, but at least one major objection would be removed. Supersymmetry might well end up being true, whether or not string theory as a whole is shown to accurately describe nature.

Dark energy: The discovery string theory should have predicted

Astronomers found evidence in 1998 that the expansion of the universe was actually accelerating. This accelerated expansion is caused by the *dark energy* that appears so often in the news. Not only did string theory not predict the existence of dark energy, but attempts to use science's best theories to calculate the amount of dark energy comes up with a number that's vastly larger than the one observed by astronomers. The theory just absolutely failed to initially make sense of dark energy.

Claiming this as a flaw of string theory is a bit more controversial than the other two, but there's some (albeit questionable) logic behind it. The goal of string theory is nothing less than the complete rewriting of gravitational law, so it's not unreasonable to think that string theory should have anticipated dark energy in some way. When Einstein constructed his theory of general relativity, the mathematics indicated that space could be expanding (later proved to be true). When Paul Dirac formulated a quantum theory of the electron, the mathematics indicated an antiparticle existed (later proved to actually exist). A profound theory like string theory can be expected to illuminate new facts about our universe, not be blind-sided by unanticipated discoveries.

Of course, no other theory anticipated an accelerating expansion of the universe either. Prior to the observational evidence (some of which is still contested, as you find out in Chapter 19), cosmologists (and string theorists) had no reason to assume that the expansion rate of space was increasing. Years after dark energy was discovered, it was shown that string theory could be modified to include it, which string theorists count as a success (although the critics continue to be unsatisfied).

Where did all of these "fundamental" theories come from?

Unfortunately, as string theorists performed more research, they had a growing problem (pun intended). Instead of narrowing in on a single *vacuum* (solution) that could be used to explain the universe, it began to look like there were an absurdly large number of vacua. Some physicists' hopes that a unique, fundamental version of string theory would fall out of the mathematics effectively dissolved.

In truth, such hype was rarely justified in the first place. In general relativity, for example, an infinite number of ways to solve the equations exist, and the goal is to find solutions that match our universe. The overly ambitious string theorists (the ones who expected a single vacuum to fall out of the sky) soon realized that they, too, would end up with a rich *string theory landscape,* as Leonard Susskind calls the range of possible vacua (see Chapter 11 for more on the Susskind's landscape idea). The goal of string theory has since become to figure out which set of vacua applies to our universe.

Looking into String Theory's Future

At present, string theory faces two hurdles. The first is the theoretical hurdle, which is whether a model can be formulated that describes our own universe. The second hurdle is the experimental one, because even if string theorists are successful in modeling our universe, they'll then have to figure out how to make a distinct prediction from the theory that's testable in some way.

Right now, string theory falls short on both counts, and it's unclear whether it can ever be formulated in a way that will be uniquely testable. The critics claim that growing disillusionment with string theory is rising among theoretical physicists, while the supporters continue to talk about how string theory is being used to resolve the major questions of the universe.

Only time will tell whether string theory is right or wrong, but regardless of the answer, string theory has driven scientists for years to ask fundamental questions about our universe and explore the answers to those questions in new ways. Even an alternative theory would in part owe its success to the hard work performed by string theorists.

Theoretical complications: Can we figure out string theory?

The current version of string theory is called *M-theory,* introduced in 1995, which is a comprehensive theory that includes the five supersymmetric string theories. M-Theory exists in 11 dimensions. There's just one problem. No one knows what M-theory is.

Scientists are searching for a complete string theory, but they don't have one yet. And, until they do, there's no way of knowing that they'll be successful. Until string theorists have a complete theory that describes our own universe, the theory could all be smoke and mirrors. Although some aspects of string theory may be shown to be true, it may be that these are only approximations of some more fundamental theory — or it may be that string theory is actually that fundamental theory itself.

String theory, the driving force of 21st-century theoretical physics, *could* prove to be nothing more than a mathematical illusion that provides some approximate insights into science but isn't actually the theory that drives the forces of nature.

It's unclear how long the search for a theory can last without some specific breakthrough. There's a sense (among some) that the most brilliant physicists on the planet have been spinning their wheels for decades, with only a handful of significant insights, and even those discoveries don't seem to lead anywhere specific.

The theoretical implications of string theory are addressed in Chapters 10 and 11, while the criticism of the theory rears its ugly head in Chapter 17.

Experimental complications: Can we prove string theory?

Even if a precise version of string theory (or M-theory) is formulated, the question then moves from the theoretical to the experimental realm. Right now, the energy levels that scientists can reach in experiments are probably way too small to realistically test string theory, although aspects of the theory can be tested today.

Theory moves forward with directions from experiment, but the last input that string theory had from experiment was the realization that it failed as a theory describing the scattering of particles within particle accelerators.

The realm string theory claims to explain involves distances so tiny that it's questionable whether scientists will ever achieve a technology able to probe at that length, so it's possible that string theory is an inherently untestable theory of nature. (Some versions of string theory do make predictions in testable ranges, however, and string theorists hope that these versions of string theory may apply to our universe.)

You find out some ways to possibly test string theory in Chapter 12, although these are only speculative because right now science doesn't even have a theory that makes any unique predictions. The best physicists can hope for are some hints, such as the discovery of extra dimensions of certain types, new cosmological predictions about the formation of our universe, or the missing supersymmetric particles, that would give some direction to the theoretical search.

Part II
The Physics Upon Which String Theory Is Built

The 5th Wave   By Rich Tennant

1905   At a lunch counter in Bern, Einstein formulates his Special Theory of Relishivity.

In this part . . .

String theory is an evolution of concepts that have been around for at least 300 years. To understand the theory and its implications, you have to first understand certain fundamental concepts, such as how scientific theories develop.

In this part, you see how science progresses, which will be helpful as you encounter the various scientific revolutions that have led to string theory. I introduce physics concepts at the heart of string theory, ranging from the smallest distance measurable to the entire universe. These overviews allow you to follow the later string theory topics. However, the chapters in this part don't even come close to providing complete explanations of the fundamental topics from classical physics, relativity, quantum physics, particle physics, and cosmology.

For more detailed introductions to the physics concepts addressed in Part II, I recommend *Physics For Dummies*, *Einstein For Dummies*, *Quantum Physics For Dummies*, and *Astronomy For Dummies,* 2nd edition, (all published by Wiley) as excellent starting points.

Chapter 4

Putting String Theory in Context: Understanding the Method of Science

String theory is at the cutting edge of science. It's a mathematical theory of nature that, at present, makes few predictions that are testable. This begs the question of what it takes for a theory to be scientific.

In this chapter, I look a bit more closely at the methods scientists use to investigate nature's structure. I explore how scientists perform science and some of the ways that their work is viewed. I certainly don't solve any of these big, philosophical issues in this chapter, but my goal is to make it clear that scientists have differing views about how the nature of science is supposed to work. Although I could write pages and pages on the evolution of scientific thought throughout the ages, I touch on these topics in just enough detail to help you understand some of the arguments in favor of and against string theory.

Exploring the Practice of Science

Before you can figure out whether string theory is scientific, you have to ask, "What is science?"

Science is the methodical practice of trying to understand and predict the consequences of natural phenomena. This is done through two distinct but closely related means: theory and experiment.

Not all science is created equal. Some science is performed with diagrams and mathematical equations. Other science is performed with costly experimental apparatus. Still other forms of science, while also costly, involve observing distant galaxies for clues to the mystery of the universe.

String theory has spent more than 30 years focusing on the theory side of the scientific equation and, sadly, is lacking on the experimental side, as critics never hesitate to point out. Ideally, the theories developed would eventually be validated by experimental evidence. (See the later sections "The need for experimental falsifiability" and "The foundation of theory is mathematics" for more on the necessity of experimentation.)

The myth of the scientific method

When in school, I was taught that science followed nice, simple rules called the *scientific method*. These rules are a classical model of scientific investigation based on principles of *reductionism* and *inductive logic*. In other words, you take observations, break them down (the reductionism part), and use them to create generalized laws (the inductive logic part). String theory's history certainly doesn't follow this nice classical model.

In school, the steps of the scientific method actually changed a bit depending on the textbook I had in a given year, though they generally had mostly common elements. Frequently, they were delineated as a set of bullet points:

- **Observe a phenomenon:** Look at nature
- **Formulate a hypothesis:** Ask a question (or propose an answer)
- **Test the hypothesis:** Perform an experiment
- **Analyze the data:** Confirm or reject the hypothesis

Breaking down nature with Bacon

The ideas of the scientific method are often traced back to Sir Francis Bacon's 1620 book *Novum Organum.* It proposed that reductionism and inductive reasoning could be used to arrive at fundamental truths about the causes of natural events.

In the Baconian model, the scientist breaks natural phenomena down into component parts that are then compared to other components based on common themes. These reduced categories are then analyzed using principles of inductive reasoning.

Inductive reasoning is a logical system of analysis where you start with specific true statements and work to create generalized laws, which would apply to all situations, by finding commonalities between the observed truths.

In a way, this scientific method is a myth. I earned a degree in physics, with honors no less, without once being asked a question about the scientific method in a physics course. (It did come up in my Philosophy of Science course, which you can thank for much of this chapter.)

Turns out there's no single scientific method that all scientists follow. Scientists don't look at a list and think, "Well, I've observed my phenomenon for the day. Time to formulate my hypothesis." Instead, science is a dynamic activity that involves a continuous, active analysis of the world. It's an interplay between the world we observe and the world we conceptualize. Science is a translation between observations, experimental evidence, and the hypotheses and theoretical frameworks that are built to explain and expand on those observations.

Still, the basic ideas of the scientific method do tend to hold. They aren't so much hard and fast rules, but they're guiding principles that can be combined in different ways depending on what's being studied.

The need for experimental falsifiability

Traditionally, the idea has been that an experiment can either confirm or refute a theory. An experimental result yields *positive evidence* if it supports the theory, while a result that contradicts the hypothesis is *negative evidence.*

In the 20th century, a notion arose that the key to a theory — the thing that makes it scientific — is whether it can in some way be shown to be false. This *principle of falsifiability* can be controversial when applied to string theory, which theoretically explores energy levels that can't at present (or possibly ever) be directly explored experimentally. Some claim that because string theory currently fails the test of falsifiability, it's somehow not "real science." (Check out Chapter 17 for more on this idea.)

The focus on this falsifiability is traced back to philosopher Karl Popper's 1934 book *The Logic of Scientific Discovery*. He was opposed to the reductionist and inductive methods that Francis Bacon had popularized three centuries earlier. In a time that was characterized by the rise of modern physics, it appeared that the old rules no longer applied.

Popper reasoned that the principles of physics arose not merely by viewing little chunks of information, but by creating theories that were tested and repeatedly failed to be proved false. Observation alone could not have led to these insights, he claimed, if they'd never been put in positions to be proven false. In the most extreme form, this emphasis on falsifiability states that scientific theories don't tell you anything definite about the world, but are only the best guesses about the future based on past experience.

For example, if I predict that the sun will rise every morning, I can test this by looking out my window every morning for 50 days. If the sun is there every day, I have not proved that the sun will be there on the 51st day. After I actually observe it on the 51st day, I'll know that my prediction worked out again, but I haven't proved anything about the 52nd day, the 53rd, and so on.

No matter how good a scientific prediction is, if you can run a test that shows that it's false, you have to throw out the idea (or, at least, modify your theory to explain the new data). This led the 19th century biologist Thomas Henry Huxley to define the great tragedy of science as "the slaying of a beautiful hypothesis by an ugly fact."

To Popper, this was far from tragic, but was instead the brilliance of science. The defining component of a scientific theory, the thing that separates it from mere speculation, is that it makes a falsifiable claim.

Popper's claim is sometimes controversial, especially when being used by one scientist (or philosopher) to discredit an entire field of science. Many still believe that reduction and inductive reasoning can, in fact, lead to the creation of meaningful theoretical frameworks that represent reality as it is, even if there's no claim that is falsifiable.

String theory founder Leonard Susskind makes just this argument. He believes not in falsification, but rather in *confirmation* — you can have direct positive evidence for a theory, rather than just a lack of negative evidence against it.

This viewpoint comes out of an online debate between Susskind and physicist Lee Smolin (you can view the debate at www.edge.org/3rd_culture/ smolin_susskind04/smolin_susskind.html). In the debate, Susskind lists several examples of theories that have been denounced as unfalsifiable: behaviorism in psychology along with quark models and inflationary theory in physics.

The examples he provides are cases where scientists believe that certain traits couldn't be examined and methods were later developed that allowed them to be tested. There's a difference between being unable to falsify a theory in practice and being unable to falsify it in principle.

It may seem as if this debate over confirmation and falsifiability is academic. That's probably true, but some physicists see string theory as a battle over the very meaning of physics. Many string theory critics believe that it's inherently unfalsifiable, while string theorists believe a mechanism to test (and falsify) the prediction of string theory will be found.

The foundation of theory is mathematics

In physics, complex mathematical models are built that represent the underlying physical laws that nature follows. These mathematical models are the real theories of physics that physicists can then relate to meaningful events in the real world through experiment and other means.

Science requires both experiment and theory to build explanations of what happens in the world. To paraphrase Einstein, science without theory is lame, while science without experiment is blind.

If physics is built on a foundation of experimental observation, then theoretical physics is the blueprint that explains how those observations fit together. The insights of theory have to move beyond the details of specific observations and connect them in new ways. Ideally, these connections lead to other predictions that are testable by experiment. String theory has not yet made this significant leap from theory to experiment.

A large part of the work in theoretical physics is developing mathematical models — frequently including simplifications that aren't necessarily realistic — that can be used to predict the results of future experiments. When physicists "observe" a particle, they're really looking at data that contains a set of numbers that they have interpreted as having certain characteristics. When they look into the heavens, they receive energy readings that fit certain parameters and explanations. To a physicist, these aren't "just" numbers; they're clues to understanding the universe.

High-energy physics (which includes string theory and other physics at high energies) has an intense interplay between theoretical insights and experimental observations. Research papers in this area fall into one of four categories:

- ✔ Experiment
- ✔ Lattice (computer simulations)
- ✔ Phenomenology
- ✔ Theory

Phenomenology is the study of phenomena (no one ever said physicists were creative when it comes to naming conventions) and relating them within the framework of an existing theory. In other words, scientists focus on taking the existing theory and applying it to the existing facts or build models describing anticipated facts that may be discovered soon. Then they make predictions about what experimental observations should be obtained. (Of course, phenomenology has a lot more to it, but this is the basics of what you need to know to understand it in relation to string theory.) It's an intriguing discipline, and one that has, in recent years, begun to focus on supersymmetry and string theory. When I discuss how to possibly test string theory in Chapter 12, it is largely the work of phenomenologists that tells scientists what they're looking for.

Though scientific research can be conducted with these different methods, there is certainly overlap. Phenomenologists can work on pure theory and can also, of course, prepare a computer simulation. Also, in some ways, a computer simulation can be viewed as a process that is both experimental and theoretical. But what all of these approaches have in common is that the scientific results are expressed in the language of science: mathematics.

The rule of simplicity

In science, one goal is to propose the fewest "entities" or rules needed to explain how something works. In many ways, the history of science is seen as a progression of simplifying the complex array of natural laws into fewer and fewer fundamental laws.

Take *Occam's razor,* which is a principle developed in the 14th century by Franciscan friar and logician William of Occam. His "law of parsimony" is basically translated (from Latin) as "entities must not be multiplied beyond necessity." (In other words, keep it simple.) Albert Einstein famously stated a similar rule as "Make everything as simple as possible, but not simpler." Though not a scientific law itself, Occam's razor tends to guide how scientists formulate their theories.

In some ways, string theory seems to violate Occam's razor. For example, in order for string theory to work, it requires the addition of a lot of odd components (extra dimensions, new particles, and other features mentioned in Chapters 10 and 11) that scientists haven't actually observed yet. However, if these components are indeed necessary, then string theory is in accord with Occam's razor.

The role of objectivity in science

Some people believe that science is purely objective. And, of course, science *is* objective in the sense that the principles of science can be applied in the same way by anyone and get the same results. But the idea that scientists are themselves inherently objective is a nice thought, but it's about as true as the notion of pure objectivity in journalism. The debate over string theory demonstrates that the discussion isn't always purely objective. At its core, the debate is over different opinions about how to view science.

In truth, scientists make choices continually that are subjective, such as which questions to pursue. For example, when string theory founder Leonard Susskind met Nobel Prize winner Murray Gell-Mann, Gell-Mann laughed at the very idea of vibrating strings. Two years later, Gell-Mann wanted to hear more about it.

In other words, physicists are people. They have learned a difficult discipline, but this doesn't make them infallible or immune to pride, passion, or any other human foible. The motivation for their decisions may be financial, aesthetic, personal, or any other reason that influences human decisions.

The degree to which a scientist relies on theory versus experiment in guiding his activities is another subjective choice. Einstein, for example, spoke of the ways in which only the "free inventions of the mind" (pure physical principles, conceived in the mind and aided by the precise application of mathematics) could be used to perceive the deeper truths of nature in ways that pure experiment never could. Of course, had experiments never confirmed his "free inventions," it's unlikely that I or anyone else would be citing him a century later.

Understanding How Scientific Change Is Viewed

The debates over string theory represent fundamental differences in how to view science. As the first part of this chapter points out, many people have proposed ideas about what the goals of science should be. But over the years, science changes as new ideas are introduced, and it's in trying to understand the nature of these changes where the meaning of science really comes into question.

The methods in which scientists adapt old ideas and adopt new ones can also be viewed in different ways, and string theory is all about adapting old ideas and adopting new ones.

Old becomes new again: Science as revolution

The interplay between experiment and theory is never so obvious as in those realms where they fail to match up. At that point, unless the experiment contained a flaw, scientists have no choice but to adapt the existing theory to fit the new evidence. The old theory must transform into a new theory. The philosopher of science Thomas Kuhn spoke of such transformations as *scientific revolutions.*

In Kuhn's model (which not all scientists agree with), science progresses along until it accumulates a number of experimental problems that make scientists redefine the theories that science operates under. These overarching theories are *scientific paradigms,* and the transition from one paradigm to a new one is a period of upheaval in science. In this view, string theory would be a new scientific paradigm, and physicists would be in the middle of the scientific revolution where it gains dominance.

A scientific paradigm, as proposed by Kuhn in his 1962 *The Structure of Scientific Revolutions,* is a period of business as usual for science. A theory explains how nature works, and scientists work within this framework.

Kuhn views the Baconian scientific method — regular puzzle-solving activities — as taking place within an existing scientific paradigm. The scientist gains facts and uses the rules of the scientific paradigm to explain them.

The problem is that there always seems to be a handful of facts that the scientific paradigm can't explain. A few pieces of data don't seem to fit. During the periods of normal science, scientists do their best to explain this data, to incorporate it into the existing framework, but they aren't overly concerned about these occasional anomalies.

That's fine when there are only a few such problems, but when enough of them pile up, it can pose serious problems for the prevailing theory.

As these abnormalities begin to accumulate, the activity of normal science becomes disrupted and eventually reaches the point where a full scientific revolution takes place. In a *scientific revolution,* the current scientific paradigm is replaced by a new one that offers a different conceptual model of how nature functions.

At some point, scientists can't just proceed with business as usual anymore, and they're forced to look for new ways to interpret the data. Initially, scientists attempt to do this with minor modifications to the existing theory. They can tack on an exception here or a special case there. But if there are enough anomalies, and if these makeshift fixes don't resolve all the problems, scientists are forced to build a new theoretical framework.

In other words, they are forced not only to amend their theory, but to construct an entirely new paradigm. It isn't just that some factual details were wrong, but their most basic assumptions were wrong. In a period of scientific revolution, scientists begin to question everything they thought they knew about nature. For example, in Chapter 10 you see that string theorists have been forced to question the number of dimensions in the universe.

Combining forces: Science as unification

Science can be seen as a progressive series of *unifications* between ideas that were, at one point, seen as separate and distinct. For example, biochemistry came about by applying the study of chemistry to systems in biology. Together with zoology, this yields genetics and *neo-darwinism* — the modern theory of evolution by natural selection, the cornerstone of biology.

In this way, we know that all biological systems are fundamentally chemical systems. And all chemical systems, in turn, come from combining different atoms to form molecules that ultimately follow the assorted laws defined in the Standard Model of particle physics.

Physics, because it studies the most fundamental aspects of nature, is the science most interested in these principles of unification. String theory, if successful, might unify all fundamental physical forces of the universe down to one single equation.

Galileo and Newton unified the heavens and Earth in their work in astronomy, defining the motion of heavenly bodies and firmly establishing that Earth followed exactly the same rules as all other bodies in our solar system. Michael Faraday and James Clerk Maxwell unified the concepts of electricity and magnetism into a single concept governed by uniform laws — electromagnetism. (If you want more information on gravity or electromagnetism, you'll be attracted to Chapter 5.)

Albert Einstein, with the help of his old teacher Hermann Minkowski, unified the notions of space and time as dimensions of space-time, through his theory of special relativity. In the same year, as part of the same theory, he unified the concepts of mass and energy as well. Years later, in his general theory of relativity, he unified gravitational force and special relativity into one theory.

Central to quantum physics is the notion that particles and waves aren't the separate phenomena that they appear to be. Instead, particles and waves can be seen as the same unified phenomenon, viewed differently in different circumstances.

The unification continued in the Standard Model of particle physics, when electromagnetism was ultimately unified with the strong and weak nuclear forces into a single framework.

This process of unification has been astoundingly successful, because nearly everything in nature can be traced back to the Standard Model — except for gravity. String theory, if successful, will be the ultimate unification theory, finally bringing gravity into harmony with the other forces.

What happens when you break it? Science as symmetry

A *symmetry* exists when you can take something, transform it in some way, and nothing seems to change about the situation. The principle of symmetry is crucial to the study of physics and has special implications for string theory in particular. When a transformation to the system causes a change in the situation, scientists say that it represents a *broken symmetry*.

This is obvious in geometry. Take a circle and draw a line through its center, as in Figure 4-1. Now picture flipping the circle around that line. The resulting image is identical to the original image when flipped about the line. This is *linear* or *reflection symmetry*. If you were to spin the figure 180 degrees, you'd end up with the same image again. This is *rotational symmetry*. The trapezoid, on the other hand, has *asymmetry* (or lacks symmetry) because no rotation or reflection of the shape will yield the original shape.

The most fundamental form of symmetry in physics is the idea of *translational symmetry,* which is where you take an object and move it from one location in space to another. If I move from one location to another, the laws of physics should be the same in both places. This principle is how scientists use laws discovered on Earth to study the distant universe.

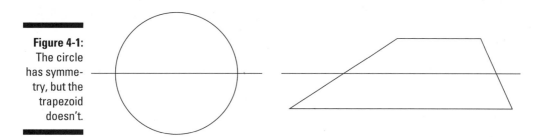

Figure 4-1:
The circle
has symme-
try, but the
trapezoid
doesn't.

In physics, though, symmetry means way more than just taking an object and flipping, spinning, or sliding it through space.

The most detailed studies of energy in the universe indicate that, no matter which direction you look, space is basically the same in all directions. The universe itself seems to have been symmetric from the very beginning.

The laws of physics don't change over time (at least according to most physicists and certainly not on short timescales, like a human lifetime or the entire age of the United States of America). If I perform an experiment today and perform the same experiment tomorrow, I'll get essentially the same result. The laws of physics possess a basic symmetry with respect to time. Changing the time of something doesn't change the behavior of the system, though I discuss some potential exceptions in Chapter 16.

These and other symmetries are seen as central to the study of science, and in fact, many physicists have stated that symmetry is the single most important concept for physics to grasp.

The truth is that while physicists often speak of the elegance of symmetry in the universe, the string theorist Leonard Susskind is quite right when he points out that things get interesting when the symmetry breaks.

In fact, as I was preparing for this book, the 2008 Nobel Prize in Physics was awarded to three physicists — Yoichiro Nambu, Makoto Kobayashi, and Toshihide Maskawa — for work in broken symmetry performed decades ago.

Without broken symmetry, everything would be absolutely uniform everywhere. The very fact that we have a chemistry that allows us to exist is proof that some aspects of symmetry don't hold up in the universe.

Many theoretical physicists believe that a symmetry exists between the four fundamental forces (gravity, electromagnetism, weak nuclear force, strong nuclear force), a symmetry that broke early in the universe's formation and causes the differences we see today. String theory is the primary (if not the only) means of understanding that broken symmetry, if it does (or did) indeed exist.

This broken symmetry may be closely linked to supersymmetry, which is necessary for string theory to become viable. Supersymmetry has been investigated in many areas of theoretical physics, even though there's no direct experimental evidence for it, because it ensures that the theory includes many desirable properties.

Supersymmetry and the unification of forces are at the heart of the string theory story. As you read more about string theory, it's up to you to determine whether the lack of experimental evidence condemns it from the start.

Chapter 5

What You Must Know about Classical Physics

*N*o matter how complex modern physics concepts get, they have their roots in basic classical concepts. To understand the revolutions leading up to string theory, you need to first understand these basic concepts. You'll then be able to understand how string theory recovers and generalizes them.

In this chapter, I present some physics concepts that you need to be familiar with to understand string theory. First, I discuss three fundamental concepts in physics: matter, energy, and how they interact. Next I explain waves and vibrations, which are crucial to understanding stringy behavior. Gravity is also key, so Sir Isaac Newton's key discoveries come next. Finally, I give a brief overview of electromagnetic radiation, an important aspect of physics that leads directly into the discovery of both relativity and quantum physics — the two theories that together give birth to modern string theory!

This Crazy Little Thing Called Physics

Physics is the study of matter and its interactions. Physics tries to understand the behavior of physical systems from the most fundamental laws that we can achieve. String theory could provide the most fundamental law and explain all of the universe in a single mathematical equation and physical theory.

One other key principle of physics is the idea that many of the laws that work in one location also work in another location — a principle known as *symmetry* (I cover this in more detail later in this section and also in Chapter 4). This connection between physics in different locations is just one sort of symmetry, allowing physics concepts to be related to each other. Science has progressed by taking diverse concepts and unifying them into cohesive physical laws.

This is a very broad definition of physics, but then physics is the broadest science. Because everything you see, hear, smell, touch, taste, or in any way interact with is made of matter and interacts according to some sort of rules, that means that physics is literally the study of anything that happens. In a way, chemistry and all the other sciences are approximations of the fundamental laws of physics.

Even if string theory (or some other "theory of everything") were to be found, there would still be need for other sciences. Trying to figure out every single physical system from string theory would be as absurd as trying to study the weather by analyzing every single atom in the atmosphere.

No laughing matter: What we're made of

One of the traits of matter (the "stuff" that everything is made of) is that it requires force to do something. (There are some exceptions to this, but as a rule a *force* is any influence that produces a change, or prevents a change, in a physical quantity.) *Mass* is the property that allows matter to resist a change in motion (in other words, the ability to resist force). Another key trait of matter is that it's *conserved,* meaning it can't be created or destroyed, but can only change forms. (Einstein's relativity showed this wasn't entirely true, as you see in Chapter 6.)

Without matter, the universe would be a pretty boring place. Matter is all around you. The book you're reading, as you lean back comfortably in your matter-laden chair, is made of matter. You yourself are made of matter. But what, exactly, is this stuff called matter?

Early philosophers and scientists try to understand matter

The question of matter's meaning dates back to at least the Greeks and Chinese philosophers, who wondered what made one thing different from another. Greek and Chinese thinkers noticed similar trends, and each devised a system for categorizing matter into five fundamental elements based on these common traits.

In ancient China, the five elements were metal, wood, water, fire, and earth. Eastern religion and philosophy used these elements and the different ways they interact to explain not only the natural world but also the moral realm.

Among the Greek philosophers, Aristotle is the most popular to have discussed their version of the five elements: fire, earth, air, water, and aether. *Aether* was supposedly an unearthly, spiritual substance that filled the universe. In this view of matter, the realm outside of Earth was composed of this aether and didn't undergo change the way our world does.

On Earth, material objects were seen as combinations of the basic elements. For example, mud was a combination of water and earth. A cloud was a combination of air and water. Lava was a combination of earth and fire.

In the 17th century, scientists' understanding of matter started to change as astronomers and physicists began to realize that the same laws govern matter both on Earth and in space. The universe isn't composed of eternal, unchanging, unearthly aether, but of hard balls of ordinary matter.

Newton's key insight into the study of matter was that it resisted change in motion (I explain this in more detail in the later "Force, mass, and acceleration: Putting objects into motion" section). The degree to which an object resists this change in motion is its *mass.*

Scientists discover that mass can't be destroyed

Antoine-Laurent Lavoisier's work in the 18th century provided physics with another great insight into matter. Lavoisier and his wife, Marie Anne, performed extensive experiments that indicated that matter can't be destroyed; it merely changes from one form to another. This principle is called the *conservation of mass.*

This isn't an obvious property. If you burn a log, when you look at the pile of ash, it certainly looks like you have a lot less matter than you started with. But, indeed, Lavoisier found that if you're extremely careful that you don't misplace any of the pieces — including the pieces that normally float away during the act of burning — you end up with as much mass at the end of the burning as you started with.

Over and over again, Lavoisier showed this unexpected trait of matter to be the case, so much so that we now take it for granted as a familiar part of our universe. Water may boil from liquid into gas, but the particles of water continue to exist and can, if care is taken, be reconstituted back into liquid. Matter can change form, but can't be destroyed (at least not until nuclear reactions, which weren't discovered until well after Lavoisier's time).

As the study of matter progresses through time, things grow stranger instead of more familiar. In Chapter 8, I discuss the modern understanding of matter, in which we are composed mostly of tiny particles that are linked together with invisible forces across vast (from their scale) empty distances. In fact, as string theory suggests, it's possible that even those tiny particles aren't really there — at least not in the way we normally picture them.

Add a little energy: Why stuff happens

The matter in our universe would never do anything interesting if it weren't for the addition of energy. There would be no change from hot to cold or from fast to slow. Energy too is conserved, as discovered through the 1800s as the laws of thermodynamics were explored, but the story of energy's conservation is more elusive than that of matter. You can see matter, but tracking energy proves to be trickier.

Kinetic energy is the energy involved when an object is in motion. *Potential energy* is the energy contained within an object, waiting to be turned into kinetic energy. It turns out that the *total* energy — kinetic energy plus potential energy — is conserved any time a physical system undergoes a change.

String theory makes predictions about physical systems that contain a *large* amount of energy, packed into a very small space. The energies needed for string theory predictions are so large that it might never be possible to construct a device able to generate that much energy and test the predictions.

The energy of motion: Kinetic energy

Kinetic energy is most obvious in the case of large objects, but it's true at all size levels. (I mean large objects in comparison to particles, so a grain of sand and the planet both would be considered large in this case.) Heat (or *thermal energy*) is really just a bunch of atoms moving rapidly, representing a form of kinetic energy. When water is heated, the particles accelerate until they break free of the bonds with other water molecules and become a gas. The motion of particles can cause energy to emit in different forms, such as when a burning piece of coal glows white hot.

Sound is another form of kinetic energy. If two billiard balls collide, the particles in the air will be forced to move, resulting in a noise. All around us, particles in motion are responsible for what takes place in our universe.

Stored energy: Potential energy

Potential energy, on the other hand, is stored energy. Potential energy takes a lot more forms than kinetic energy and can be a bit trickier to understand.

A spring, for example, has potential energy when it's stretched out or compressed. When the spring is released, the potential energy transforms into kinetic energy as the spring moves into its least energetic length.

Moving an object in a gravitational field changes the amount of potential energy stored in it. A penny held out from the top of the Empire State Building has a great deal of potential energy due to gravity, which turns into a great deal of kinetic energy when dropped (although not, as evidenced on an episode of *MythBusters,* enough to kill an unsuspecting pedestrian on impact).

This may sound a bit odd, talking about something having more or less energy just because of where it is, but the environment is part of the physical system described by the physics equations. These equations tell exactly how much potential energy is stored in different physical systems, and they can be used to determine outcomes when the potential energy gets released.

Symmetry: Why some laws were made to be broken

A change in location or position that retains the properties of the system is called a *geometric symmetry* (or sometimes *translational symmetry*). Another form of symmetry is an *internal symmetry,* which is when something within the system can be swapped for something else and the system (as a whole) doesn't change. When a symmetrical situation at high energy collapses into a lower energy *ground state* that is asymmetrical, it's called *spontaneous symmetry breaking.* An example would be when a roulette wheel spins and slows into a "ground state." The ball ultimately settles into one slot in the wheel — and the gambler either wins or loses.

String theory goes beyond the symmetries we observe to predict even more symmetries that aren't observed in nature. It predicts a necessary symmetry that's not observed in nature, called *supersymmetry.* At the energies we observe, supersymmetry is an example of a broken symmetry, though physicists believe that in high-energy situations, the supersymmetry would no longer be broken (which is what makes it so interesting to study). I cover supersymmetry in Chapters 2 and 10.

Translational symmetry: Same system, different spot

If an object has *translational symmetry,* you can move it and it continues to look the same (for a detailed explanation of this, flip to Chapter 4). Moving objects in space doesn't change the physical properties of the system.

Now, didn't I just say in the last section that the potential energy due to gravity changes depending on where an object is? Yes, I did. Moving an object's location in space can have an impact on the physical system, but the laws of physics themselves don't change (so far as we can tell). If the Empire State Building, Earth, and the penny held over the edge (the entire "system" in this example) were all shifted by the same amount in the same direction, there would be no noticeable change to the system.

Internal symmetry: The system changes, but the outcome stays the same

In an *internal symmetry*, some property of the system can undergo a change without changing the outcome of the result.

For example, changing every particle with its antiparticle — changing positive charges to negative and negative charges to positive — leaves the electromagnetic forces involved completely identical. This is a form of internal symmetry, called *charge conjugation symmetry.* Most internal symmetries aren't perfect symmetries, meaning that they behave somewhat differently in some situations.

Spontaneous symmetry breaking: A gradual breakdown

Physicists believe that the laws of the universe used to be even more symmetric, but have gone through a process called *spontaneous symmetry breaking,* where the symmetry falls apart in the universe we observe.

If everything were perfectly symmetric, the universe would be a very boring place. The slight differences in the universe — the broken symmetries — are what make the natural world so interesting, but when physicists look at the physical laws, they tend to find that the differences are fairly small in comparison to the similarities.

To understand spontaneous symmetry breaking, consider a pencil perfectly balanced on its tip. The pencil is in a state of perfect balance, of equilibrium, but it's unstable. Any tiny disturbance will cause it to fall over. However, no law of physics says *which way* the pencil will fall. The situation is perfectly symmetrical because all directions are equal.

As soon as the pencil starts to fall, however, definite laws of physics dictate the direction it will continue to fall. The symmetrical situation spontaneously (and, for all intents and purposes, randomly) begins to collapse into one definite, asymmetrical form. As the system collapses, the other options are no longer available to the system.

The Standard Model of particle physics, as well as string theory (which includes the Standard Model as a low-energy approximation), predicts that some properties of the universe were once highly symmetrical but have undergone spontaneous symmetry breaking into the universe we observe now.

All Shook Up: Waves and Vibrations

In string theory, the most fundamental objects are tiny strings of energy that vibrate or oscillate in regular patterns. In physics, such systems are called *harmonic oscillators,* and much work has been done to study them.

Though the strings of string theory are different, understanding the vibrations of classical objects — like air, water, jump-ropes, springs — can help you understand the behavior of these exotic little creatures when you encounter them. These classical objects can carry what are called *mechanical waves.*

Catching the wave

Waves (as we usually think of them) move through some sort of medium. If you flick the end of a jump-rope or string, a wave moves along the rope or string. Waves move through the water, or sound waves through the air, with those materials acting as the medium for the wave motion.

In classical physics, waves transport energy, but not matter, from one region to another. One set of water molecules transfers its energy to the nearby water molecules, which means that the wave moves through the water, even though the actual water molecules don't actually travel all the way from the start of the wave to the end of the wave.

This is even more obvious if I were to take the end of a jump-rope and shake it, causing a wave to travel along its length. Clearly, the molecules at my end of the jump-rope aren't traveling along it. Each group of jump-rope molecules is nudging the next group of jump-rope molecules, and the end result is the wave motion along its length.

There are two types of mechanical waves, as shown in Figure 5-1:

- ✔ **Transverse wave:** A wave in which the displacement of the medium is perpendicular to the direction of travel of the wave along the medium, like the flicking of a jump-rope.

- ✔ **Longitudinal wave:** A wave that moves in the same direction in which the wave travels, like a piston pushing on a cylinder of water.

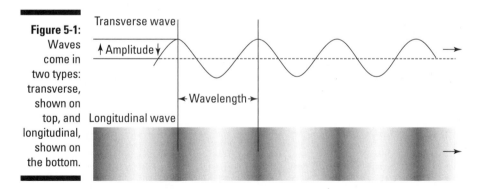

Figure 5-1: Waves come in two types: transverse, shown on top, and longitudinal, shown on the bottom.

The highest point on a transverse wave (or the densest point in a longitudinal wave) is called a *crest*. The lowest point on a transverse wave (or the least dense point in a longitudinal wave) is called a *trough*.

The displacement from the resting point to the crest — in other words, how high the wave gets — is called the *amplitude*. The distance from one crest to another (or one trough to another) is called the *wavelength.* These values are shown on the transverse wave in Figure 5-1. The wavelength is shown on the longitudinal wave, as well, although the amplitude is hard to show on that type of wave, so it isn't included.

Another useful thing to consider is the *velocity* (speed and direction) of the wave. This can be determined by its wavelength and *frequency,* which is a measure of how many times the wave passes a given point per second. If you know the frequency and the wavelength, you can calculate the velocity. This, in turn, allows you to calculate the energy contained within the wave.

Another trait of many waves is the *principle of superposition,* which states that when two waves overlap, the total displacement is the sum of the individual displacements, as shown in Figure 5-2. This property is also referred to as *wave interference.*

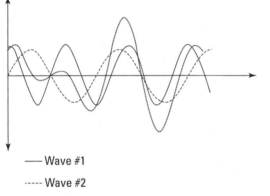

Figure 5-2: When two waves overlap, the total displacement is the sum of the two individual displacements.

—— Wave #1

----- Wave #2

········ Addition of Wave #1 and Wave #2

Consider waves when two ships cross each other's path. The waves made by the ships cause the water to become choppier, and as the waves add height to each other, they cause massive swells.

Similarly, sometimes waves can cancel each other out. If the crest of wave 1 overlaps with the trough of wave 2, they cancel each other out at that point. This sort of interference plays a key role in one of the quantum physics problems I discuss in Chapter 7 — the double slit experiment.

Getting some good vibrations

String theory depicts strings of energy that vibrate, but the strings are so tiny that you never perceive the vibrations directly, only their consequences. To understand these vibrations, you have to understand a classical type of wave called a *standing wave* — a wave that doesn't appear to be moving.

In a standing wave, certain points, called *nodes,* don't appear to move at all. Other points, called *antinodes,* have the maximum displacement. The arrangement of nodes and antinodes determines the properties of various types of standing waves.

The simplest example of a standing wave is one with a node on each end, such as a string that's fixed in place on the ends and plucked. When there is a node on each end and only one antinode in between them, the wave is said to vibrate at the *fundamental frequency.*

Consider a jump-rope that is held at each end by a child. The ends of the rope represent nodes because they don't move much. The center of the rope is the antinode, where the displacement is the greatest and where another child will attempt to jump in. This is vibration at the fundamental frequency, as demonstrated in Figure 5-3a.

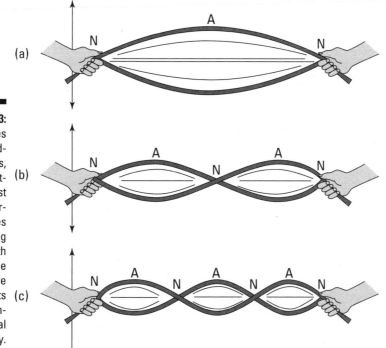

Figure 5-3:
Examples of standing waves, demonstrating the first three normal modes of a string fixed at both ends. The top wave represents the fundamental frequency.

If the children get ambitious, however, and begin putting more energy into the wave motion of their jump-rope, a curious thing happens. Eventually, the children will pump enough energy into the rope that instead of one large anti-node, two smaller antinodes are created, and the center of the rope seems to be at rest, as shown in Figure 5-3b. It's almost as if someone grabbed hold of the middle of the rope and gingerly, but firmly, is holding it in place!

A second type of standing wave can be considered if instead of a child holding each end of the rope, one end is mounted on a ring around a pole. The child holding one end begins the wave motion, but the end on the pole is now unconstrained and moves up and down. Instead of having a node on each end, one end is a node (held by the child) and the other is an antinode (moving up and down on the pole).

A similar situation in music happens when using a pipe that's closed at one end and open at the other, such as in an organ. A node forms at the closed end of the pipe, but the open end of the pipe is always an antinode.

A third type of standing wave has an antinode at each end. This would represent either a pipe that's open on both ends or a rope that's free to move on both ends.

The more energy that's pumped into the standing wave, the more nodes form (see Figure 5-3c). The series of frequencies that cause new nodes to form are called *harmonics*. (In music, harmonics are called *overtones*.) The waves that correspond to harmonics are called *normal modes,* or *vibrational modes*.

Music works because of the manipulation and superposition of harmonic overtones created by these normal modes of vibration. The first three normal modes are shown in Figure 5-3, where a string is fixed on both ends.

In string theory, the vibrational modes of strings (and other objects) are similar to those that I'm talking about in this chapter. In fact, matter itself is seen as the manifestation of standing waves on strings. Different vibrational modes give rise to different particles! We perceive the particles from the lowest vibrational modes, but with higher energies, we may be able to detect other, higher-energy particles.

Newton's Revolution: How Physics Was Born

Many see Sir Isaac Newton's discoveries as the start of modern physics (along with a bit of help from his predecessor, Galileo Galilei). Newton's discoveries dominated two centuries of physics, until Albert Einstein took his place at the apex of scientific greatness.

Newton's accomplishments are diverse, but he's known largely for four crucial discoveries that define the realm of physics even today:

- Three laws of motion
- Law of universal gravitation
- Optics
- Calculus

Each of these discoveries has elements that will prove important as you attempt to understand the later discoveries of string theory.

Force, mass, and acceleration: Putting objects into motion

Newton formulated three laws of motion, which showed his understanding of the real meaning of motion and how it relates to force. Under his laws of motion, a force created a proportional acceleration on an object.

This understanding was a necessary foundation upon which his law of gravity was built (see the next section). In fact, both were introduced in his 1686 book *Philosophiae Naturalis Principia Mathematica,* a title that translates into *Mathematical Principles of Natural Philosophy.* This book has become known by the shorter title *Principia* in physics circles.

The second law of motion says that the force required to accelerate an object is the product of the mass and acceleration, expressed by the equation $F = ma$, where F is the total force, m is the object's mass, and a is the acceleration. To figure out the total acceleration on an object, you figure out the total forces acting on it and then divide by the mass.

Strictly speaking, Newton said that force was equal to the change in momentum of an object. In calculus, this is the derivative of momentum with respect to time. Momentum is equal to mass times velocity. Because mass is assumed to be constant and the derivative of velocity with respect to time yields the acceleration, the popular $F = ma$ equation is a simplified way of looking at this situation.

This equation can also be used to define mass. If I take a force and divide it by the acceleration it causes on an object, I can determine the mass of the object. One question which string theorists hope to answer is *why* some objects have mass and others (such as the photon) do not.

Newton makes some laws about motion

The second law, and the way it relates force, acceleration, and mass, is the only law of motion relevant to a string theory discussion. However, for true Newton-o-philes, here are the other two laws of motion, paraphrased for ease of understanding:

✔ **Newton's first law of motion:** An object at rest remains at rest, or an object in motion remains in motion, unless acted upon by an external force. In other words, it takes a force to cause motion to change.

✔ **Newton's third law of motion:** When two objects interact through a force, each object exerts a force on the other object that is equal and opposite. In other words, if I exert a force on the wall with my hand, the wall exerts an equal force back on my hand.

Gravity: A great discovery

With the laws of motion in hand, Newton was able to perform the action that would make him the greatest physicist of his age: explaining the motion of the heavens and the Earth. His proposal was the *law of universal gravitation,* which defines a force acting between two objects based on their masses and the distance separating them.

The more massive the objects, the higher the gravitational force is. The relationship with distance is an *inverse* relationship, meaning that as the distance increases, the force drops off. (It actually drops off with the square of the distance — so it drops off very quickly as objects are separated.) The closer two objects are, the higher the gravitational force is.

The strength of the gravitational force determines a value in Newton's equation, called the *gravitational constant* or Newton's constant. This value is obtained by performing experiments and observations, and calculating what the constant should be. One question still open to physics and string theory is why gravity is so weak compared to other forces.

Gravity seems fairly straightforward, but it actually causes quite a few problems for physicists, because it won't behave itself and get along with the other forces of the universe. Newton himself wasn't comfortable with the idea of a force acting at a distance, without understanding the mechanism involved. But the equations, even without a thorough explanation for what caused it, worked. In fact, the equations worked well enough that for more than two centuries, until Einstein, no one could figure out what was missing from the theory. More on this in Chapter 6.

Optics: Shedding light on light's properties

Newton also performed extensive work in understanding the properties of light, a field known as *optics*. Newton supported a view that light moved as tiny particles, as opposed to a theory that light traveled as a wave. Newton performed all of his work in optics assuming that light moved as tiny balls of energy flying through the air.

For nearly a century, Newton's view of light as particles dominated, until Thomas Young's experiments in the early 1800s demonstrated that light exhibited the properties of waves, namely the principle of superposition (see the earlier "Catching the wave" section for more on superposition and the later "Light as a wave: The ether theory" section for more on light waves).

The understanding of light, which began with Newton, would lead to the revolutions in physics by Albert Einstein and, ultimately, to the ideas at the heart of string theory. In string theory, both gravity and light are caused by the behavior of strings.

Calculus and mathematics: Enhancing scientific understanding

To study the physical world, Newton had to develop new mathematical tools. One of the tools he developed was a type of math that we call *calculus*. Actually, at the same time he invented it, philosopher and mathematician Gottfried Leibniz had also created calculus completely independently! Newton needed calculus to perform his analysis of the natural world. Leibniz, on the other hand, developed it mainly to explain certain geometric problems.

Think for a moment how amazing this really is. A purely mathematical construct, like calculus, provided key insights into the physical systems that Newton explored. Alternately, the physical analysis that Newton performed led him to create calculus. In other words, this is a case where mathematics and science seemed to help build upon each other! One of the major successes of string theory is that it has provided motivation for important mathematical developments that have gone on to be useful in other realms.

The Forces of Light: Electricity and Magnetism

In the 19th century, the physical understanding of the nature of light changed completely. Experiments began to show strong cases where light acted like waves instead of particles, which contradicted Newton (see the "Optics: Shedding light on light's properties" section for more on Newton's findings). During the same time, experiments into electricity and magnetism began to reveal that these forces behaved like light, except that we couldn't see them!

By the end of the 19th century, it became clear that electricity and magnetism were different manifestations of the same force: *electromagnetism*. One of the goals of string theory is to develop a single theory that incorporates both electromagnetism and gravity.

Light as a wave: The ether theory

Newton had treated light as particles, but experiments in the 19th century began to show that light acted like a wave. The major problem with this was that waves require a medium. Something has to do the waving. Light seemed to travel through empty space, which contained no substance at all. What was the medium that light used to move through? What was waving?

To explain the problem, physicists proposed that space was filled with a substance. When looking for a name for this hypothetical substance, physicists turned back to Aristotle and named it *luminous ether*. (Some physicists continued to spell it *aether,* but I call it *ether* to distinguish it from Aristotle's fifth element.)

Even with this hypothetical ether, though, there were still problems. Newton's optics still worked, and his theory described light in terms of tiny balls moving in straight lines, not as waves! It seemed that sometimes light acted like a wave and sometimes it acted as a particle.

Most physicists of the 19th century believed in the wave theory, largely because the study of electricity and magnetism helped support the idea that light was a wave, but they were unable to find solid evidence of the ether.

Invisible lines of force: Electric and magnetic fields

Electricity is the study of how charged particles affect each other. *Magnetism,* on the other hand, is the study of how magnetized objects affect each other. In the 19th century, research began to show that these two seemingly separate phenomena were, in fact, different aspects of the same thing. The physicist Michael Faraday proposed that invisible fields transmitted the force.

Electricity and magnetism are linked together

An electrical force acts between two objects that contain a property called *electrical charge* that can be either positive or negative. Positive charges repel other positive charges, and negative charges repel other negative charges, but positive and negative charges *attract* each other, as in Figure 5-4.

Coulomb's Law, which describes the simplest behavior of the electric force between charged particles (a field called *electrostatics*), is an inverse square law, similar to Newton's law of gravity. This provided some of the first inklings that gravity and electrostatic forces (and, ultimately, electromagnetism) might have something in common.

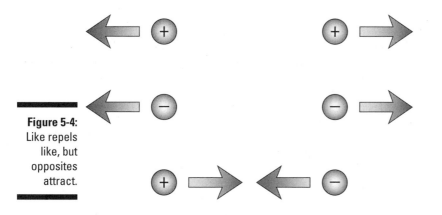

Figure 5-4:
Like repels
like, but
opposites
attract.

When electrical charges move, they create an electrical current. These currents can influence each other through a magnetic force. This was discovered by Hans Christian Oersted, who found that a wire with an electrical current running through it could deflect the needle of a compass.

Later work by Michael Faraday and others showed that this worked the other way, as well — a magnetic force can influence an electrical current. As demonstrated in Figure 5-5, moving a magnet toward a conducting loop of wire causes a current to run through the wire.

Figure 5-5:
A magnet
moving
toward a
metal ring
creates a
current in
the ring.

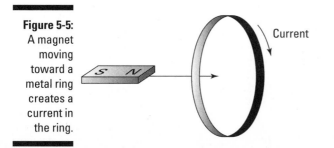

Current

Faraday proposes force fields to explain these forces

In the 1840s, Michael Faraday proposed the idea that invisible lines of force were at work in electrical currents and magnetism. These hypothetical lines made up a *force field* that had a certain value and direction at any given point and could be used to calculate the total force acting on a particle at that point. This concept was quickly adapted to also apply to gravity in the form of a *gravitational field*.

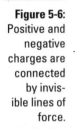

These invisible lines of force were responsible for the electrical force (as shown in Figure 5-6) and magnetic force (as shown in Figure 5-7). They resulted in an electric field and a magnetic field that could be measured.

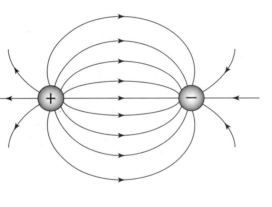

Figure 5-6:
Positive and negative charges are connected by invisible lines of force.

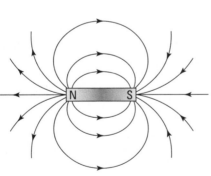

Figure 5-7:
The north and south poles of a bar magnet are connected by invisible lines of force.

Faraday proposed the invisible lines of force, but wasn't nearly as clear on how the force was transmitted, which drew ridicule from his peers. Keep in mind, though, that Newton also couldn't fully explain how gravity was transmitted, so there was precedent to this. Action at a distance was already an established part of physics, and Faraday, at least, was proposing a physical model of how it could take place.

The fields proposed by Faraday turned out to have applications beyond electricity and magnetism. Gravity, too, could be written in a field form. The benefit of a force field is that every point in space has a value and direction associated with it. If you can calculate the value of the field at a point, you know exactly how the force will act on an object placed at that point. Today, every law of physics can be written in the form of fields.

Maxwell's equations bring it all together: Electromagnetic waves

Physicists now know that electricity and magnetism are both aspects of the same *electromagnetic force*. This force travels in the form of *electromagnetic waves*. We see a certain range of this electromagnetic energy in the form of visible light, but there are other forms, such as X-rays and microwaves, that we don't see.

In the mid-1800s, James Clerk Maxwell took the work of Faraday and others and created a set of equations, known as Maxwell's equations, that described the forces of electricity and magnetism in term of *electromagnetic waves*. An electromagnetic wave is shown in Figure 5-8.

Figure 5-8:
The electric field and magnetic field are in step in an electromagnetic wave.

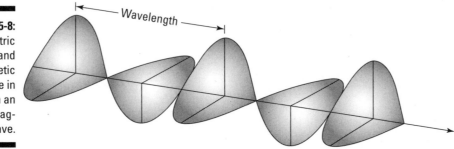

Maxwell's equations allowed him to calculate the exact speed that the electromagnetic wave traveled. When Maxwell performed this calculation, he was amazed to find that he recognized the value. Electromagnetic waves moved at exactly the speed of light!

Maxwell's equations showed that visible light and electromagnetic waves are different manifestations of the same underlying phenomena. In other words, we see only a small range of the entire spectrum of electromagnetic waves that exist in our universe. Extending this unification to include all the forces of nature, including gravity, would ultimately lead to theories of quantum gravity such as string theory.

Two dark clouds and the birth of modern physics

Two significant unanswered questions with the electromagnetic theory remained. The first problem was that the ether hadn't been detected, while the second involved an obscure problem about energy radiation, called the blackbody problem (described in Chapter 7). What's amazing, in retrospect, is that physicists didn't see these problems (or *dark clouds* as British scientist Lord Kelvin called them in a 1900 speech) as especially significant, but instead believed that they were minor issues that would soon be resolved. As you see in Chapters 6 and 7, resolving these two problems would introduce the great revolutions of modern physics — relativity and quantum physics.

Chapter 6

Revolutionizing Space and Time: Einstein's Relativity

* *

* *

Albert Einstein introduced his theory of relativity to explain the issues arising from the electromagnetic concepts introduced in Chapter 5. The theory has had far-reaching implications, altering our understanding of time and space. It provides a theoretical framework that tells us how gravity works, but it has left open certain questions that string theory hopes to answer.

In this book, I give you only a glimpse of relativity — the glimpse needed to understand string theory. For a more in-depth look at the fascinating concepts of Einstein's theory of relativity, I suggest *Einstein For Dummies* by Carlos I. Calle, PhD (Wiley).

In this chapter, I explain how the ether model failed to match experimental results and how Einstein introduced special relativity to resolve the problem. I discuss Einstein's theory of gravity in general relativity, including a brief look at a rival theory of gravity and how Einstein's theory was confirmed. I then point out some issues arising from relativity. Finally, I introduce a theory that tried to unify relativity and electromagnetics and is seen by many as a predecessor of string theory.

What Waves Light Waves? Searching for the Ether

In the latter part of the 19th century, physicists were searching for the mysterious *ether* — the medium they believed existed for light waves to wave through. Their inability to discover this ether, despite good experiments, was frustrating, to say the least. Their failure paved the way for Einstein's explanation, in the form of the theory of relativity.

As I explain in Chapter 5, waves had to pass through a medium, a substance that actually did the waving. Light waves pass through "empty space" of a *vacuum* (a space without any air or other regular matter), so physicists had predicted a luminous ether that must exist everywhere and be some sort of substance that scientists had never before encountered. In other words, the "empty space" was not (in the view of the time) really empty because it contained ether.

Some things could be predicted about the ether, though. For example, if there was a medium for light, the light was moving through it, like a swimmer moving through the water. And, like a swimmer, the light should travel slightly faster when going in the same direction as the water's current than when the swimmer is trying to go against the water's current.

This doesn't mean that the ether itself was moving. Even if the ether was completely still, Earth was moving within the ether, which is effectively the same thing. If you walk through a still body of water, it feels basically the same as if you were walking in place and the water was flowing around you. (In fact, they now have small pools that use this exact principle. You can swim for hours in a pool that's only a few feet long. Because a powerful current is pumping through it, you swim against the current and never go anywhere.)

Physicists wanted to construct an experiment based on this concept that would test whether light traveled different speeds in different directions. This sort of variation would support the idea that light was traveling through an ether medium.

In 1881, physicist Albert Michelson created a device called an *interferometer* designed to do just that. With the help of his colleague Edward Morley, he improved the design and precision of the device in 1887. The Michelson-Morley interferometer is shown in Figure 6-1.

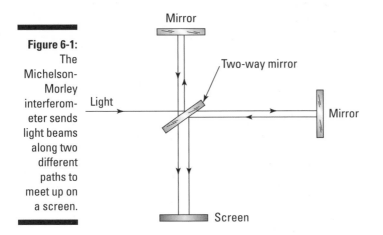

Figure 6-1:
The Michelson-Morley interferometer sends light beams along two different paths to meet up on a screen.

The interferometer used mirrors that were only partially reflective, so they let half the light pass through and reflected half the light. The interferometer set these mirrors at an angle, splitting a single beam of light so it ended up traveling two different paths. The paths traveled perpendicular to each other, but ended up hitting the same screen.

In 1887, Michelson and Morley ran a series of tests with the improved interferometer to discover the ether. They thought that the light traveling along one of these paths should be slightly faster than the light traveling along the other path, because one of them would be going either with or against the ether, and the other path would be perpendicular to the ether. When the light hit the screen, each beam would have traveled the exact same distance. If one had traveled a slightly different speed, the two beams would be slightly out of phase with each other, which would show distinctive wave interference patterns — light and dark bands would appear — on the screen.

No matter how many times Michelson and Morley conducted the experiment, they never found this difference in speed for the two light beams. They always found the same speed of approximately 670 million miles per hour, regardless of the direction the light traveled.

Physicists didn't immediately dismiss the ether model; instead they (including Michelson and Morley) considered it a failed experiment, even though it *should* have worked had there been an ether. In 1900, when Lord Kelvin gave his "two dark clouds" speech, 13 years had passed without being able to detect the ether's motion, but it was still assumed that the ether existed.

Sometimes scientists are reluctant to give up on a theory that they've devoted years to, even if the evidence turns against them — something that the critics of string theory believe may be happening right now in the theoretical physics community.

No Ether? No Problem: Introducing Special Relativity

In 1905, Albert Einstein published a paper explaining how to have electromagnetics work without an ether. This theory came to be known as the *theory of special relativity,* which explains how to interpret motion between different *inertial frames of reference* — that is, places that are moving at constant speeds relative to each other.

The key to special relativity was that Einstein explained the laws of physics when two objects are moving at a constant speed as the *relative motion* between the two objects, instead of appealing to the ether as an absolute frame of reference that defined what was going on. If you and some astronaut, Amber, are moving in different spaceships and want to compare your observations, all that matters is how fast you and Amber are moving with respect to each other.

Special relativity includes only the special case (hence the name) where the motion is uniform. The motion it explains is only if you're traveling in a straight line at a constant speed. As soon as you accelerate or curve — or do anything that changes the nature of the motion in any way — special relativity ceases to apply. That's where Einstein's general theory of relativity comes in, because it can explain the general case of any sort of motion. (I cover this theory later in the chapter.)

Einstein's 1905 paper that introduced special relativity, "On the Electrodynamics of Moving Bodies," was based on two key principles:

- **The principle of relativity:** The laws of physics don't change, even for objects moving in inertial (constant speed) frames of reference.

- **The principle of the speed of light:** The speed of light is the same for all observers, regardless of their motion relative to the light source. (Physicists write this speed using the symbol c.)

The genius of Einstein's discoveries is that he looked at the experiments and assumed the findings were true. This was the exact opposite of what other physicists seemed to be doing. Instead of assuming the theory was correct and that the experiments failed, he assumed that the experiments were correct and the theory had failed.

The ether had caused a mess of things, in Einstein's view, by introducing a medium that caused certain laws of physics to work differently depending on how the observer moved relative to the ether. Einstein just removed the ether entirely and assumed that the laws of physics, including the speed of light equal to c, worked the same way regardless of how you were moving — exactly as experiments and mathematics showed them to be!

Giving credit where credit is due

No physicist works in a vacuum, and that was certainly true of Albert Einstein. Though he revolutionized the world of physics, he did so by resolving the biggest issues of his day, which means he was tackling problems that a lot of other physicists were also working on. He had a lot of useful research to borrow from. Some have accused Einstein of plagiarism, or implied that his work wasn't truly revolutionary because he borrowed so heavily from the work of others.

For example, his work in special relativity was largely based on the work of Hendrik Lorentz, George FitzGerald, and Jules Henri Poincaré, who had developed mathematical transformations that Einstein would later use in his theory of relativity. Essentially, they did the heavy lifting of creating special relativity, but they fell short in one important way — they thought it was a mathematical trick, not a true representation of physical reality.

The same is true of the discovery of the photon. Max Planck introduced the idea of energy in discrete packets, but thought it was only a mathematical trick to resolve a specific odd situation. Einstein took the mathematical results literally and created the theory of the photon.

The accusations of plagiarism are largely dismissed by the scientific community because Einstein never denied that the work was done by others and, in fact, gave them credit when he was aware of their work. Physicists tend to recognize the revolutionary nature of Einstein's work and know that others contributed greatly to it.

Unifying space and time

Einstein's theory of special relativity created a fundamental link between space and time. The universe can be viewed as having three space dimensions — up/down, left/right, forward/backward — and one time dimension. This 4-dimensional space is referred to as the *space-time continuum*.

If you move fast enough through space, the observations that you make about space and time differ somewhat from the observations that other people, who are moving at different speeds, make. The formulas Einstein used to describe these changes were developed by Hendrik Lorentz (see the nearby sidebar, "Giving credit where credit is due").

String theory introduces many more space dimensions, so grasping how the dimensions in relativity work is a crucial starting point to understanding some of the confusing aspects of string theory. The extra dimensions are so important to string theory that they get their own chapter, Chapter 13.

Following the bouncing beam of light

The reason for this space-time link comes from applying the principles of relativity and the speed of light very carefully. The speed of light is the distance light travels divided by the time it takes to travel this path, and (according to Einstein's second principle) all observers must agree on this speed.

Sometimes, though, different observers disagree on the distance a light beam has traveled, depending on how they're moving through space.

This means that to get the same speed those observers must *disagree* about the time the light beam travels the given distance.

You can picture this for yourself by understanding the thought experiment depicted in Figure 6-2. Imagine that you're on a spaceship and holding a laser so it shoots a beam of light directly up, striking a mirror you've placed on the ceiling. The light beam then comes back down and strikes a detector.

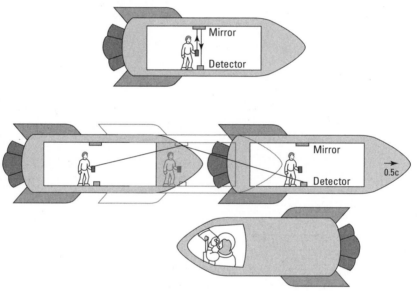

Figure 6-2:
(Top) You see a beam of light go up, bounce off the mirror, and come straight down. (Bottom) Amber sees the beam travel along a diagonal path.

However, the spaceship is traveling at a constant speed of half the speed of light (0.5c, as physicists would write it). According to Einstein, this makes no difference to you — you can't even tell that you're moving. However, if astronaut Amber were spying on you, as in the bottom of Figure 6-2, it would be a different story.

Amber would see your beam of light travel upward along a diagonal path, strike the mirror, and then travel downward along a diagonal path before striking the detector. In other words, you and Amber would see *different* paths for the light and, more importantly, those paths aren't even the same length. This means that the time the beam takes to go from the laser to the mirror to the detector must also be different for you and Amber so that you both agree on the speed of light.

This phenomenon is known as *time dilation,* where the time on a ship moving very quickly appears to pass slower than on Earth. In Chapter 16, I explain some ways that this aspect of relativity can be used to allow time travel. In fact, it allows the only form of time travel that scientists know for sure is physically possible.

As strange as it seems, this example (and many others) demonstrates that in Einstein's theory of relativity, space and time are intimately linked together. If you apply Lorentz transformation equations, they work out so that the speed of light is perfectly consistent for both observers.

This strange behavior of space and time is only evident when you're traveling close to the speed of light, so no one had ever observed it before. Experiments carried out since Einstein's discovery have confirmed that it's true — time and space are perceived differently, in precisely the way Einstein described, for objects moving near the speed of light.

Building the space-time continuum

Einstein's work had shown the connection between space and time. In fact, his theory of special relativity allows the universe to be shown as a 4-dimensional model — three space dimensions and one time dimension. In this model, any object's path through the universe can be described by its *worldline* through the four dimensions.

Though the concept of space-time is inherent in Einstein's work, it was actually an old professor of his, Hermann Minkowski, who developed the concept into a full, elegant mathematical model of space-time coordinates in 1907. Actually, Minkowski had been specifically unimpressed with Einstein, famously calling him a "lazy dog."

One of the elements of this work is the Minkowski diagram, which shows the path of an object through space-time. It shows an object on a graph, where one axis is space (all three dimensions are treated as one dimension for simplicity) and the other axis is time. As an object moves through the universe, its sequence of positions represents a line or curve on the graph, depending on how it travels. This path is called the object's *worldline,* as shown in Figure 6-3. In string theory, the idea of a worldline becomes expanded to include the motion of strings, into objects called *worldsheets.* (See Chapter 16 for more information. A worldsheet can be seen in Figure 16-1.)

Unifying mass and energy

The most famous work of Einstein's life also dates from 1905 (a very busy year for him), when he applied the ideas of his relativity paper to come up with the equation $E=mc^2$ that represents the relationship between mass *(m)* and energy *(E)*.

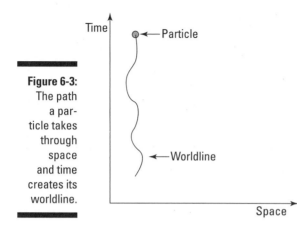

Figure 6-3:
The path
a par-
ticle takes
through
space
and time
creates its
worldline.

The reason for this connection is a bit involved, but essentially it relates to the concept of kinetic energy discussed in Chapter 5. Einstein found that as an object approached the speed of light, *c,* the mass of the object increased. The object goes faster, but it also gets heavier. In fact, if it were actually able to move at *c,* the object's mass and energy would both be infinite. A heavier object is harder to speed up, so it's impossible to ever actually get the particle up to a speed of *c.*

In this 1905 paper — "Does the Inertia of a Body Depend on its Energy Content?" — Einstein showed this work and extended it to stationary matter, showing that mass at rest contains an amount of energy equal to mass times c^2.

Until Einstein, the concepts of mass and energy were viewed as completely separate. He proved that the principles of conservation of mass and conservation of energy are part of the same larger, unified principle, *conservation of mass-energy.* Matter can be turned into energy and energy can be turned into matter because a fundamental connection exists between the two types of substance.

If you're interested in greater detail on the relationship of mass and energy, check out *Einstein For Dummies* (Wiley) or the book *E=mc²: A Biography of the World's Most Famous Equation* by David Bodanis (Walker & Company).

Changing Course: Introducing General Relativity

General relativity was Einstein's theory of gravity, published in 1915, which extended special relativity to take into account *non-inertial frames of reference* — areas that are accelerating with respect to each other. General relativity takes the form of field equations, describing the curvature of space-time and the distribution of matter throughout space-time. The effects of matter and space-time on each other are what we perceive as gravity.

Gravity as acceleration

Einstein immediately realized that his theory of special relativity worked only when an object moved in a straight line at a constant speed. What about when one of the spaceships accelerated or traveled in a curve?

Einstein came to realize the principle that would prove crucial to developing his general theory of relativity. He called it the *principle of equivalence,* and it states that an accelerated system is completely physically equivalent to a system inside a gravitational field.

As Einstein later related the discovery, he was sitting in a chair thinking about the problem when he realized that if someone fell from the roof of a house, he wouldn't feel his own weight. This suddenly gave him an understanding of the equivalence principle.

As with most of Einstein's major insights, he introduced the idea as a thought experiment. If a group of scientists were in an accelerating spaceship and performed a series of experiments, they would get exactly the same results as if sitting still on a planet whose gravity provided that same acceleration, as shown in Figure 6-4.

Einstein's brilliance was that after he realized an idea applied to reality, he applied it uniformly to every physics situation he could think of.

For example, if a beam of light entered an accelerating spaceship, then the beam would appear to curve slightly, as in the left picture of Figure 6-5. The beam is trying to go straight, but the ship is accelerating, so the path, as viewed inside the ship, would be a curve.

By the principle of equivalence, this meant that gravity should also bend light, as shown in the right picture of Figure 6-5. When Einstein first realized this in 1907, he had no way to calculate the effect, other than to predict that it would probably be very small. Ultimately, though, this exact effect would be the one used to give general relativity its strongest support.

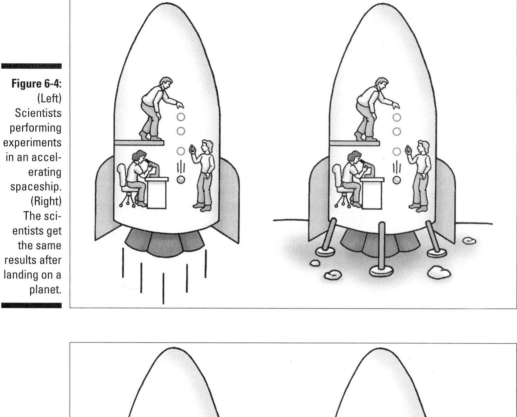

Figure 6-4:
(Left) Scientists performing experiments in an accelerating spaceship. (Right) The scientists get the same results after landing on a planet.

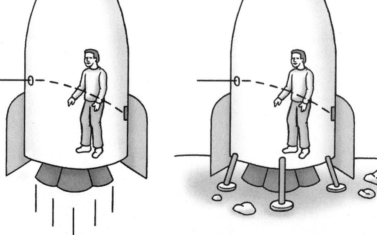

Figure 6-5:
Both acceleration and gravity bend a beam of light.

Gravity as geometry

The theory of the space-time continuum already existed, but under general relativity Einstein was able to describe gravity as the bending of space-time geometry. Einstein defined a set of *field equations,* which represented the way that gravity behaved in response to matter in space-time. These field equations could be used to represent the geometry of space-time that was at the heart of the theory of general relativity.

As Einstein developed his general theory of relativity, he had to refine Minkowski's notion of the space-time continuum into a more precise mathematical framework (see the earlier "Building the space-time continuum" section for more on this concept). He also introduced another principle, *the principle of covariance.* This principle states that the laws of physics must take the same form in all coordinate systems.

In other words, all space-time coordinates are treated the same by the laws of physics — in the form of Einstein's field equations. This is similar to the relativity principle, which states that the laws of physics are the same for all observers moving at constant speeds. In fact, after general relativity was developed, it was clear that the principles of special relativity were a special case.

Einstein's basic principle was that no matter where you are — Toledo, Mount Everest, Jupiter, or the Andromeda galaxy — the same laws apply. This time, though, the laws were the field equations, and your motion could very definitely impact what solutions came out of the field equations.

Applying the principle of covariance meant that the space-time coordinates in a gravitational field had to work exactly the same way as the space-time coordinates on a spaceship that was accelerating. If you're accelerating through empty space (where the space-time field is flat, as in the left picture of Figure 6-6), the geometry of space-time would appear to curve. This meant that if there's an object with mass generating a gravitational field, it had to curve the space-time field as well (as shown in the right picture of Figure 6-6).

Figure 6-6:
Without matter, space-time is flat (left), but it curves when matter is present (right).

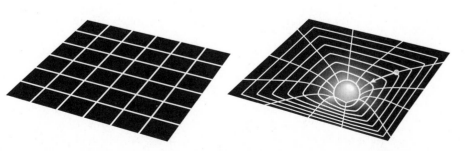

In other words, Einstein had succeeded in explaining the Newtonian mystery of where gravity came from! Gravity resulted from massive objects bending space-time geometry itself.

Because space-time curved, the objects moving through space would follow the "straightest" path along the curve, which explains the motion of the planets. They follow a curved path around the sun because the sun bends space-time around it.

Again, you can think of this by analogy. If you're flying by plane on Earth, you follow a path that curves around the Earth. In fact, if you take a flat map and draw a straight line between the start and end points of a trip, that would *not* be the shortest path to follow. The shortest path is actually the one formed by a "great circle" that you'd get if you cut the Earth directly in half, with both points along the outside of the cut. Traveling from New York City to northern Australia involves flying up along southern Canada and Alaska — nowhere close to a straight line on the flat maps we're used to.

Similarly, the planets in the solar system follow the shortest paths — those that require the least amount of energy — and that results in the motion we observe.

Testing general relativity

For most purposes, the theory of general relativity matched the predictions of Newton's gravity, and it also incorporated special relativity — it was a relativistic theory of gravity. But no matter how impressive a theory is, it still has to be confirmed by experiment before the physics community fully embraces it. Today, scientists have seen extensive evidence of general relativity.

One stunning modern example of applying relativity is the global positioning system (GPS). The GPS satellite system sends carefully synchronized beams around the planet. This is what allows military and commercial devices to know their location to within a few meters or better. But the entire system is based upon the synchronization of these satellites that had to be programmed with corrections to take into account the curvature of space-time near Earth. Without these corrections, minor timing errors would accumulate day after day, causing the system to completely break down.

Of course, such equipment wasn't available to Einstein when he published his theory in 1915, so the theory had to gain support in other ways.

One solution that Einstein immediately arrived at was to explain an anomaly in the orbit of Mercury. For years, it had been known that Newtonian gravity wasn't quite matching up with astronomers' observations of Mercury's path around the sun. By taking into account the effects of relativity's curved space-time, Einstein's solution precisely matched the path observed by astronomers.

Still, this wasn't quite enough to win over all the critics, because another theory of gravity had its own appeal.

Pulled in another direction: Einstein's competition for a theory of gravity

A couple of years before Einstein completed his theory of general relativity, the Finnish physicist Gunnar Nordström introduced his metric theory of gravity that also combined gravity with special relativity. He went further, taking James Clerk Maxwell's electromagnetic theory and applying an extra space dimension, which meant that the electromagnetic force was also included in the theory. It was simpler and more comprehensive than Einstein's general relativity, but ultimately wrong (in a way that most physicists then and today see as fairly obvious). But this was the first attempt to use an extra dimension in a unification theory, so it's worth investigating a bit.

Einstein himself was supportive of Nordström's work to incorporate special relativity with gravity. In a 1913 speech on the state of unifying the two, he said that only his work and that of Nordström met the necessary criteria. In 1914, though, Nordström introduced a mathematical trick that increased the stakes of unification. He took Maxwell's electromagnetic equations and formulated them in four space dimensions, instead of the usual three that Einstein had used. The resulting equations included the equation describing the force of gravity!

Including the dimension of time, this made Nordström's theory a 5-dimensional space-time theory of gravity. He treated our universe as a 4-dimensional projection of a 5-dimensional space-time. (This is kind of similar to how your shadow on a wall is a 2-dimensional projection of your 3-dimensional body.) By adding an extra dimension to an established physical theory, Nordström unified electromagnetics and gravity! This provides an early example of a principle from string theory — that the addition of extra dimensions can provide a mathematical means for unifying and simplifying physical laws.

When Einstein published his complete theory of general relativity in 1915, Nordström jumped ship on his own theory because Einstein could explain Mercury's orbit while his own theory could not.

Nordström's theory had a lot going for it, though, because it was much simpler than Einstein's theory of gravity. In 1917, a year after Nordström himself had given up on it, some physicists considered his metric theory a valid alternative to general relativity. Nothing noteworthy came out of these scientists' efforts, though, so clearly they had backed the wrong theory.

The eclipse that confirmed Einstein's life work

One major difference between Einstein's and Nordström's theories was that they made different predictions about light's behavior. Under Nordström's theory, light always traveled in a straight line. According to general relativity, a beam of light would curve within a gravitational field.

In fact, as early as the late 1700s, physicists had predicted that light would curve under Newtonian gravity. Einstein's equations showed that these earlier predictions were off by a factor of 2.

The deflection of light predicted by Einstein is due to the curvature of space-time around the sun. Because the sun is so massive that it causes space-time to curve, a beam of light that travels near the sun will travel along a curved path — the "shortest" path along the curved space-time, as shown in Figure 6-7.

Figure 6-7:
Light from distant stars follows the shortest path along curved space-time, according to Einstein's theory of general relativity.

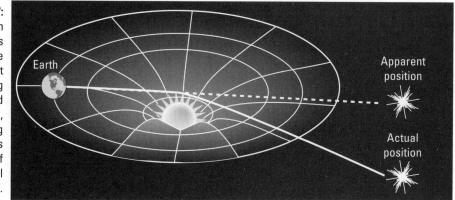

In 1911, Einstein had done enough work on general relativity to predict how much the light should curve in this situation, which should be visible to astronomers during an eclipse.

Astronomers on an expedition to Russia in 1914 attempted to observe the deflection of light by the sun, but the team ran into one little snag: World War I. Arrested as prisoners of war and released a few weeks later, the astronomers missed the eclipse that would have tested Einstein's theory of gravity.

This turned out to be great news for Einstein, because his 1911 calculations contained an error! Had the astronomers been able to view the eclipse in 1914, the negative results might have caused Einstein to give up his work on general relativity.

When he published his complete theory of general relativity in 1915, he'd corrected the problem, making a slightly modified prediction for how the light would be deflected. In 1919, another expedition set out, this time to the west African island of Principe. The expedition leader was British astronomer Arthur Eddington, a strong supporter of Einstein.

Despite hardships on the expedition, Eddington returned to England with the pictures he needed, and his calculations showed that the deflection of light precisely matched Einstein's predictions. General relativity had made a prediction that matched observation.

Albert Einstein had successfully created a theory that explained the gravitational forces of the universe and had done so by applying a handful of basic principles. To the degree possible, the work had been confirmed, and most of the physics world agreed with it.

Almost overnight, Einstein's name became world famous. In 1921, Einstein traveled through the United States to a media circus that probably wasn't matched until the Beatlemania of the 1960s.

Applying Einstein's Work to the Mysteries of the Universe

Einstein's work in developing the theory of relativity had shown amazing results, unifying key concepts and clarifying important symmetries in the universe. Still, there are some cases where relativity predicts strange behavior, such as *singularities,* where the curvature of space-time becomes infinite and the laws of relativity seem to break down. String theory today continues this work by trying to extend the concepts of relativity into these areas, hoping to find new rules that work in these regions.

With relativity in place, physicists could look to the heavens and begin a study of how the universe evolved over time, a field called *cosmology*. However, Einstein's field equations also allow for some strange behavior — such as black holes and time travel — that has caused great distress to Einstein and others over the years.

If you haven't read about relativity before, this chapter may seem like a whirlwind of strange, exotic concepts — and these new theories certainly felt so to the physicists of the time. Fundamental concepts — motion, mass, energy, space, time, and gravity — were transformed in a period of only 15 years!

Motion, instead of being just some incidental behavior of objects, was now crucial to understanding how the laws of physics manifested themselves. The laws don't change — this was key to all of Einstein's work — but they can manifest in different ways, depending on where you are and how you're moving — or how space-time is moving around you.

In Chapter 9, I cover the ideas of modern cosmology arising from Einstein's work, such as the black holes that can form when massive quantities of mass cause space-time to curve infinitely far and similar problems that come up when trying to apply relativity to the early universe. Or, as you see in Chapter 16, some solutions to Einstein's equations allow time travel.

Einstein himself was extremely uncomfortable with these unusual solutions to his equations. To the best of his ability, he tried to disprove them. When he failed, he would sometimes violate his own basic belief in the mathematics and claim that these solutions represented physically impossible situations.

Despite the strange implications, Einstein's theory of general relativity has been around for nearly a century and has met every challenge — at least when applied to objects larger than a molecule. As I point out in Chapter 2, at very small scales quantum effects become important, and the description using general relativity begins to break down. The equations make no sense, and space-time becomes an exotic, tumultuous mess of energy fluctuations. The force of gravity explodes to an infinite value. String theory (hopefully) represents one way of reconciling gravity at this realm, as I explain in Chapters 10 and 11.

Kaluza-Klein Theory — String Theory's Predecessor

One of the earliest attempts to unify gravity and electromagnetic forces came in the form of *Kaluza-Klein theory,* a short-lived theory that again unified the forces by introducing an extra space dimension. In this theory, the extra space dimension was curled up to a microscopic size. Though it failed, many of the same concepts were eventually applied in the study of string theory.

Einstein's theory had proved so elegant in explaining gravity that physicists wanted to apply it to the other force known at the time — the electromagnetic force. Was it possible that this other force was also a manifestation of the geometry of space-time?

In 1915, even before Einstein completed his general relativity field equations, the British mathematician David Hilbert said that research by Nordström and others indicated "that gravitation and electrodynamics are not really different." Einstein responded, "I have often tortured my mind in order to bridge the gap between gravitation and electromagnetism."

One theory in this regard was developed and presented to Einstein in 1919 by German mathematician Theodor Kaluza. In 1914, Nordström had written Maxwell's equations in five dimensions and had obtained the gravity equations (see the section "Pulled in another direction: Einstein's competition for a theory of gravity"). Kaluza took the gravitational field equations of general relativity and wrote them in five dimensions, obtaining results that included Maxwell's equations of electromagnetism!

When Kaluza wrote to Einstein to present the idea, the founder of relativity replied by saying that increasing the dimensions "never dawned on me" (which means he must have been unaware of Nordström's attempt to unify electromagnetism and gravity, even though he was clearly aware of Nordström's theory of gravity).

In Kaluza's view, the universe was a 5-dimensional cylinder and our 4-dimensional world was a projection on its surface. Einstein wasn't quite ready to take that leap without any evidence for the extra dimension. Still, he incorporated some of Kaluza's concepts into his own unified field theory that he published and almost immediately recanted in 1925.

A year later, in 1926, Swedish physicist Oskar Klein dusted off Kaluza's theory and reworked it into the form that has come to be known as the *Kaluza-Klein theory*. Klein introduced the idea that the fourth space dimension was rolled up into a tiny circle, so small that there was essentially no way for us to detect it directly.

In Kaluza-Klein theory, the geometry of this extra, hidden space dimension dictated the properties of the electromagnetic force — the size of the circle, and a particle's motion in that extra dimension, related to the electrical charge of a particle. The physics fell apart on this level because the predictions of an electron's charge and mass never worked out to match the true value. Also, many physicists initially intrigued with the Kaluza-Klein theory became far more intrigued with the growing field of quantum mechanics, which had actual experimental evidence (as you see in Chapter 7).

Another problem with the theory is that it predicted a particle with zero mass, zero spin, and zero charge. Not only was this particle never observed (despite the fact that it should have been, because it's a low-energy particle), but the particle corresponded to the radius of the extra dimensions. It didn't make sense to add a theory with extra dimensions and then have a result be that the extra dimensions effectively didn't exist.

There is another (though less conventional) way to describe the failure of Kaluza-Klein theory, viewing it as a fundamental theoretical limitation: For electromagnetism to work, the extra dimension's geometry had to be completely fixed.

In this view, tacking an extra dimension onto a theory of dynamic space should result in a theory that is still dynamic. Having a fifth dimension that's fixed (while the other four dimensions are flexible) just doesn't make sense from this point of view. This concept, called *background dependence,* returns as a serious criticism of string theory in Chapter 17.

Whatever the ultimate reason for its failure, Kaluza-Klein theory lasted for only a short time, although there are indications that Einstein continued to tinker with it off and on until the early 1940s, incorporating elements into his various failed unified field theory attempts.

In the 1970s, as physicists began to realize that string theory contained extra dimensions, the original Kaluza-Klein theory served as an example from the past. Physicists once again curled up the extra dimensions, as Klein had done, so they were essentially undetectable (I explain this in more detail in Chapter 10). Such theories are called Kaluza-Klein theories.

Chapter 7

Brushing Up on Quantum Theory Basics

*A*s strange as relativity may have seemed to you (see Chapter 6), it's a cakewalk compared to understanding quantum physics. In this strange realm of physics — the realm of the extremely small — particles don't have definite positions or energies. They can exist not only as particles, but also as waves, but only when you don't look at them. One hope scientists have is that string theory will explain some of the unusual results in quantum physics or, at the least, reconcile it with general relativity. Particle physics, on the other hand, is at the heart of string theory's origins and is a direct consequence of this early work in quantum physics (see Chapter 8). Without quantum physics, string theory could not exist.

As in the other chapters in this part, the goal of this chapter is not to provide a complete overview of all of quantum physics — there are other books that do a fine job of that, including *Quantum Physics For Dummies* by Steven Holzner (Wiley). My goal here is to give you the background you need to know about quantum physics so you can understand certain aspects of string theory. It may not seem that these ideas relate directly to string theory, but being familiar with these concepts will be handy down the road when I explain string theory itself.

In this chapter, I give you a brief introduction to the history and principles of quantum physics, just enough so you can understand the later concepts related to string theory. I explain how quantum theory allows objects to act as both particles and waves. You explore the implications of the uncertainty principle and probability in quantum physics (dead cat not required). I list some of the many interpretations of what all of these strange quantum rules may actually mean — though no one really knows (or can know) for sure. Finally, I discuss the idea that special natural units can be used to describe reality.

Unlocking the First Quanta: The Birth of Quantum Physics

Quantum physics traces its roots back to 1900, when German physicist Max Planck proposed a solution to a *thermodynamics* problem — a problem having to do with heat. He resolved the problem by introducing a mathematical trick — if he assumed that energy was bundled in discrete packets, or *quanta,* the problem went away. (It proved to be brilliant because it worked. There was no theoretical reason for doing this, until Einstein came up with one five years later, as discussed in the next section.) In the process of doing this, Planck used a quantity known as *Planck's constant,* which has proved essential to quantum physics — and string theory.

Planck used this quantum concept — the concept that many physical quantities come in discrete units — to solve a problem in physics, but even Planck himself assumed that this was just a clever mathematical process to remove the infinity. It would take five years for Albert Einstein to continue the quantum revolution in physics.

The blackbody radiation problem, which Planck was trying to solve, is a basic thermodynamics problem where you have an object that is so hot that it glows inside. A small hole allows the light to escape, and it can be studied. The problem is that in the 1800s, experiments and theories in this area didn't match up.

A hot object radiates heat in the form of light (hot coals in a fire or the metal rings on electric stoves are both good examples of this). If this object were open inside, like an oven or a metal box, the heat would bounce around inside. This sort of object was called a *blackbody* — because the object itself doesn't reflect light, only radiates heat — and throughout the 1800s, various theoretical work in thermodynamics had examined the way heat behaved inside a blackbody.

Now assume that there's a small opening — like a window — in the oven, through which light can escape. Studying this light reveals information about the heat energy within the blackbody.

Essentially, the heat inside a blackbody took the form of electromagnetic waves, and because the oven is metal, they're standing waves, with nodes where they meet the side of the oven (see Chapter 5 for details about waves). This fact — along with an understanding of electromagnetics and thermodynamics — can be used to calculate the relationship between light's intensity (or brightness) and wavelength.

REMEMBER

The result is that as the wavelength of light gets very small (the ultraviolet range of electromagnetic energy), the intensity is supposed to increase dramatically, approaching infinity.

In nature, scientists never actually observe infinities, and this was no exception (see Chapter 2 for more about infinities). The research showed that there were maximum intensities in the ultraviolet range, which completely contradicted the theoretical expectations, as shown in Figure 7-1. This discrepancy came to be known as the *ultraviolet catastrophe*.

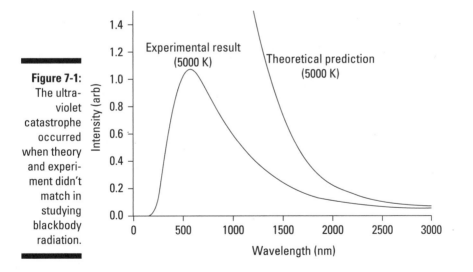

Figure 7-1: The ultra-violet catastrophe occurred when theory and experiment didn't match in studying blackbody radiation.

The ultraviolet catastrophe threatened to undermine the theories of electromagnetics and/or thermodynamics. Clearly, if they didn't match experiment, then one or both of the theories contained errors.

When Planck resolved the ultraviolet catastrophe in 1900, he did so by introducing the idea that the atom could only absorb or emit light in *quanta* (or discrete bundles of energy). One implication of this radical assumption was that there would be less radiation emitted at higher energies. By introducing the idea of discrete energy packets — by quantizing energy — Planck produced a solution that resolved the situation without having to dramatically revise the existing theories (at least at that time).

Planck's insight came when he looked at the data and tried to figure out what was going on. Clearly, the long wavelength predictions were close to matching with experiment, but the short wavelength light was not. The theory was over-predicting the amount of light that would be produced at short wavelengths, so he needed a way to limit this short wavelength.

Knowing some things about waves, Planck knew that the wavelength and frequency were inversely related. So if you're talking about waves with short wavelength, you're also talking about waves with high frequency. All he had to do was find a way to lower the amount of radiation at high frequencies.

Planck reworked the equations, assuming that the atoms could only emit or absorb energy in finite quantities. The energy and frequency were related by a proportion called *Planck's constant*. Physicists use the variable h to represent Planck's constant in his resulting physics equations.

The resulting equation worked to explain the experimental results of blackbody radiation. Planck, and apparently everyone else, thought this was just a mathematical sleight of hand that had resolved the problem in one strange, special case. Little did anyone realize that Planck had just laid the foundation for the strangest scientific discoveries in the history of the world.

Fun with Photons: Einstein's Nobel Idea of Light

Einstein received the Nobel Prize not for relativity, but instead for his work in using Planck's idea of the quantum to explain another problem — the photoelectric effect. He went further than Planck, suggesting that *all* electromagnetic energy was quantized. Light, Einstein said, moved not in waves, but in packets of energy. These packets of energy became called *photons*. Photons are one of the fundamental particles of physics that physicists hope to explain using string theory.

Powered by the photoelectric effect

Modern solar cells work off the same principle as the photoelectric effect. Composed of photoelectric materials, they take electromagnetic radiation in the form of sunlight and convert it into free electrons. Those free electrons then run through wires to create an electric current that can power devices such as ornamental lights in your flowerbed or NASA's Martian rovers.

The *photoelectric effect* occurs when light shines on certain materials that then emit electrons. It's almost as if the light knocks loose the electrons, causing them to fly off the material. The photoelectric effect was first observed in 1887 by Heinrich Hertz, but it continued to puzzle physicists until Einstein's 1905 explanation.

At first, the photoelectric effect didn't seem that hard to explain. The electrons absorbed the light's energy, which caused the electrons to fly off the metal plate. Physicists still knew very little about electrons — and virtually nothing about the atom — but this made sense.

As expected, if you increased the light's *intensity* (the total energy per second carried by the beam), more electrons definitely were emitted (see the top of Figure 7-2). There were two unexpected problems though:

- ✔ Above a certain wavelength, no electrons are emitted — no matter how intense the light is (as shown in the bottom of Figure 7-2).
- ✔ When you increase the light's intensity, the speed of the electrons doesn't change.

Einstein saw a connection between this first problem and the ultraviolet catastrophe faced by Max Planck (see the preceding section for more about Planck's work), but in the opposite direction. The longer wavelength light (or light with lower frequency) failed to do things that were being achieved by the shorter wavelength light (light with higher frequency).

Planck had created a proportional relationship between energy and frequency. Einstein again did what he was best at — he took the mathematics at face value and applied it consistently. The result was that the high frequency light had higher energy photons, so it was able to transfer enough energy into the electron for it to get knocked loose. The lower frequency photons didn't have enough energy to help any electrons escape. The photons had to have energy above a certain threshold to knock the electrons loose.

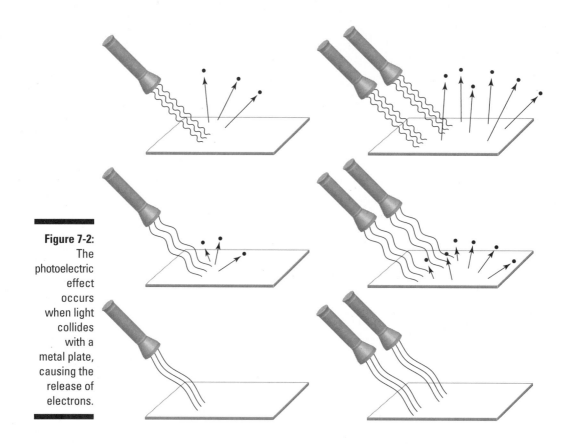

Figure 7-2:
The
photoelectric
effect
occurs
when light
collides
with a
metal plate,
causing the
release of
electrons.

Similarly, the second problem of the noneffect of light's intensity on an electron's speed is also solved by Einstein's quantum view of light. Each photon's energy is based on its frequency (or wavelength), so increasing the intensity doesn't change the energy of each photon; it only increases the total number of photons. This is why increasing the intensity causes more electrons to get emitted, but each electron maintains the same speed. The individual photon knocks out an electron with the same energy as before, but more photons are doing the same job. No single electron gets the benefit of the increase in intensity.

Based on the principle that the speed of light was constant (the basis of his special theory of relativity), Einstein knew that these photons would always move at the same velocity, c. Their energy would be proportional to the frequency of the light, based on Planck's definitions.

Waves and Particles Living Together

Within quantum physics, two alternate explanations of light work, depending on the circumstances. Sometimes light acts like a wave, and sometimes it acts like a particle, the photon. As quantum physics continued to grow, this *wave particle duality* would come up again and again, as even particles seemed to begin acting like waves. The explanation for this strange behavior lies in the *quantum wavefunction,* which describes the behavior of individual particles in a wave-like way. This strange quantum behavior of particles and waves is crucial to understanding quantum theories, such as string theory.

Einstein's theory of special relativity had seemingly destroyed the theory of an ether medium, and with his theory of the photon he proved how light could work without it. The problem was that for more than a century, there had been proof that light did, indeed, act like a wave.

Light as a wave: The double slit experiment

The experiment that proved that light acts like a wave was the *double slit experiment.* It showed a beam of light passing through two slits in a barrier, resulting in light and dark interference bands on a screen. This sort of interference is a hallmark of wave behavior, meaning that light had to be in the form of waves.

These interference patterns in light had been observed in Isaac Newton's time, in the work of Francesco Maria Grimaldi. These experiments were vastly improved upon by the young experimenter Thomas Young in 1802.

For the experiment to work, the light passing through the two slits needed to have the same wavelength. Today, you can accomplish this with lasers, but they weren't available in Young's day, so he came up with an ingenious way to get a single wavelength. He created a single slit and let light pass through that, and then that light went through two slits. Because the light passing through the two slits came from the same source, they were in phase with each other, and the experiment worked. This experimental setup is shown in Figure 7-3.

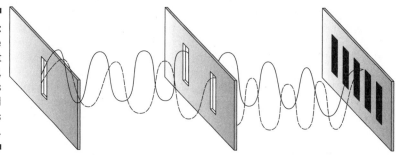

As you can see in the figure, the end result is a series of bright and dark bands on the final screen. This comes from the interference of the light waves, shown back in Figure 5-2 in Chapter 5. Recall that *interference* means you add the amplitude of the waves. Where high and low amplitudes overlap, they cancel each other out, resulting in dark bands. If high amplitudes overlap, the amplitude of the total wave is the sum of them, and the same happens with low amplitudes, resulting in the light bands.

This dual behavior was the problem facing Einstein's photon theory of light, because though the photon had a wavelength, according to Einstein, it was still a particle! How could a particle possibly have a wavelength? Conceptually, it made no real sense, until a young Frenchman offered a resolution to the situation.

Particles as a wave: The de Broglie hypothesis

In 1923, Frenchman Louis de Broglie proposed a bold new theory: Particles of matter also had wavelengths and could behave as waves, just as photons did.

Here was de Broglie's line of reasoning. Under special relativity, matter and energy were different manifestations of the same thing. The photon, a particle of energy, had a wavelength associated with it. Therefore, particles of matter, such as electrons, should also have wavelengths. His PhD dissertation set out to calculate what that wavelength (and other wave properties) should be.

Two years later, two American physicists demonstrated de Broglie's experiment by performing experiments that showed interference patterns with electrons, as shown in Figure 7-4. (The 1925 experiment wasn't actually a double slit experiment, but it showed the interference clearly. The double slit experiment with electrons was conducted in 1961.)

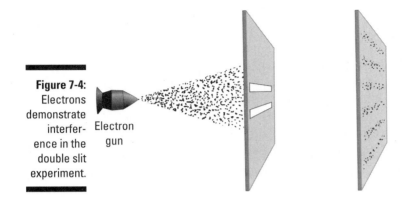

Figure 7-4:
Electrons
demonstrate
interfer-
ence in the
double slit
experiment.

Electron
gun

This behavior showed that whatever quantum law governed photons also governed particles. The wavelength of particles such as the electron is very small compared to the photon. For larger objects, the wavelength is even smaller still, quickly becoming so small as to become unnoticeable. This is why this sort of behavior doesn't show up for larger objects. If you flung baseballs through the two slits, you'd never notice an interference pattern.

Still, this left open the question of what was causing the wave behavior in these particles of energy or matter. The answer would be at the core of the new field of quantum mechanics. (String theory will later say that both types of particles — matter and energy — are manifestations of vibrating strings, but that's about 50 years down the road from de Broglie's time.)

You can picture the problem if you look at the way the experiment is set up in Figure 7-5. The light wave passes through *both* slits, and that's why the waves interfere with each other. But an electron — or a photon, for that matter — *cannot* pass through both slits at the same time if you think of them the way we're used to thinking of them; it has to pick a slit. In this classical case (where the photon is a solid object that has a certain position), there shouldn't be any interference. The beam of electrons should hit the screen in one general spot, just as if you were throwing baseballs through a hole against a wall. (This is why quantum physics challenges our classical thinking about objects and was deemed so controversial in its early years.)

In fact, if you close one of the slits, this is exactly what happens. When a slit is closed, the interference pattern goes away — the photons or electrons collect in a single band that spreads out from the brightest spot at the center.

So the interference patterns can't be explained by particles bouncing off the side of the slits or anything normal like that. It's a genuinely strange behavior that required a genuinely strange solution — in the form of quantum mechanics.

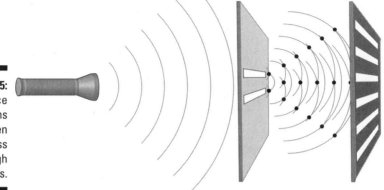

Figure 7-5:
Interference
patterns
occur when
waves pass
through
both slits.

Quantum physics to the rescue: The quantum wavefunction

The solution to the problem took the form of the *quantum wavefunction*, developed by Erwin Schrödinger. In this function, the location of the particle is dictated by a wave equation describing the probability of the particle's existence at a given point, even though the particle has a definite location when measured.

Schrödinger's wavefunction was based in part on his reading of de Broglie's hypothesis about matter having a wavelength. He used this behavior to analyze atomic models created by Niels Bohr (which I cover in Chapter 8). The resulting wavefunction explained the behavior of these atoms in terms of waves. (Bohr's student, Werner Heisenberg, had come up with a different mathematical representation to solve the atomic problem. Heisenberg's matrix method was later shown to be mathematically equivalent to Schrödinger's wavefunction. This sort of parallel work comes up often in physics, as you'll see in Chapters 10 and 11 about the development of string theory.)

The wavefunction created the wave behavior. In this viewpoint, the wave passed through both slits, even though no single, classical individual particle could pass through both slits. The wavefunction, which describes the probability of the particle arriving at a point, can be thought of as passing through both slits and creating the interference pattern. This is an interference pattern of probabilities, even though the particles themselves end up having a definite location (and therefore must pass through one slit).

Still, this isn't the end of the odd story of the double slit experiment. The strange dual behavior — wave and particle — was still there. But now a mathematical framework existed that allowed physicists to talk about the duality in a manner that made some sort of mathematical sense. The theory still held many more mysteries to be uncovered.

Why We Can't Measure It All: The Uncertainty Principle

Werner Heisenberg is best known in quantum physics for his discovery of the *uncertainty principle,* which states that the more precisely you measure one quantity, the less precisely you can know another associated quantity. The quantities sometimes come in set pairs that can't both be completely measured. One consequence of this is that to make measurements of very short distances — such as those required by string theory — very high energies are required.

What Heisenberg found was that the observation of a system in quantum mechanics disturbs the system enough that you can't know everything about the system. The more precisely you measure the position of a particle, for example, the less it's possible to precisely measure the particle's momentum. The degree of this uncertainty was related directly to Planck's constant — the same value that Max Planck had calculated in 1900 in his original quantum calculations of thermal energy. (You'll shortly see that Planck's constant has a lot of unusual implications.)

Heisenberg found that certain complementary quantities in quantum physics were linked by this sort of uncertainty:

- Position and momentum (momentum is mass times velocity)
- Energy and time

This uncertainty is a very odd and unexpected result from quantum physics. Until this time, no one had ever made any sort of prediction that knowledge was somehow inaccessible on a fundamental level. Sure, there were technological limitations to how well a measurement was made, but Heisenberg's uncertainty principle went further, saying that nature itself doesn't allow you to make measurements of both quantities beyond a certain level of precision.

One way to think about this is to imagine that you're trying to observe a particle's position very precisely. To do so, you have to look at the particle. But you want to be very precise, which means you need to use a photon with a very short wavelength, and a short wavelength relates to a high energy. If the photon with high energy hits the particle — which is exactly what you need to have happen if you want to observe the particle's position precisely — then it's going to give some of its energy to the particle. This means that any measurement you also try to make of the particle's momentum will be off. The more precisely you try to measure the position, the more you throw off your momentum measurement!

Similar explanations work if you observe the particle's momentum precisely, so you throw off the position measurement. The relationship of energy and time has a similar uncertainty. These are mathematical results that come directly out of analyzing the wavefunction and the equations de Broglie used to describe his waves of matter.

How does this uncertainty manifest in the real world? For that, let me return to your favorite quantum experiment — the double slit. The double slit experiment has continued to grow odder over the years, yielding stranger and stranger results. For example:

- ✔ If you send the photons (or electrons) through the slits one at a time, the interference pattern shows up over time (recorded on a film), even though each photon (or electron) has seemingly nothing to interfere with.

- ✔ If you set up a detector near either (or both) slits to detect which slit the photon (or electron) went through, the interference pattern goes away.

- ✔ If you set up the detector but leave it turned off, the interference pattern comes back.

- ✔ If you set up a means of determining later what slit the photon (or electron) went through, but do nothing to impact it right now, the interference pattern goes away.

What does all of this have to do with the uncertainty principle? The common denominator among the cases where the interference pattern goes away is that a measurement was made on which slit the photons (or electrons) passed through.

When no slit measurement is made, the uncertainty in position remains high, and the wave behavior appears dominant. As soon as a measurement is made, the uncertainty in position drops significantly and the wave behavior vanishes. (There is also a case where you observe *some* of the photons or electrons. Predictably, in this case, you get both behaviors, in exact ratio to how many particles you're measuring.)

Dead Cats, Live Cats, and Probability in Quantum Physics

In the traditional interpretation of quantum physics, the wavefunction is seen as a representation of the probability that a particle will be in a given location. After a measurement is made, the wavefunction collapses, giving the particle a definite value for the measured quantity.

In the double slit experiments, the wavefunction splits between the two slits, and this wavefunction results in an interference of probabilities on the screen. When the measurements are made on the screen, the probabilities are distributed so that it's more likely to find particles in some places and less likely to find them in other places, resulting in the light and dark interference bands. The particle never splits, but the probability of where the particle will be does split. Until the measurement is made, the distribution of probabilities is all that exists.

This interpretation was developed by the physicist Max Born and grew to be the core of the Copenhagen interpretation of quantum mechanics (which I explain toward the end of this chapter). For this explanation, Born received (three decades later) the 1954 Nobel Prize in Physics.

Almost as soon as the explanation of probabilities was proposed, Erwin Schrödinger came up with a morbid thought experiment intended to show how absurd it was. It's become one of the most important, and misunderstood, concepts in all of physics: the Schrödinger's cat experiment.

In this experiment, Schrödinger hypothesized a radioactive particle that has a 50 percent chance of decaying within an hour. He proposed that you place the radioactive material within a closed box next to a Geiger counter that would detect the radiation. When the Geiger counter detects the radiation from the decay, it will break a glass of poison gas. Also inside the box is a cat. If the glass breaks, the cat dies. (I told you it was morbid.)

Now, according to Born's interpretation of the wavefunction, after an hour the atom is in a quantum state where it is both decayed and not decayed — 50 percent chance of each result. This means the Geiger counter is in a state where it's both triggered and not triggered. The glass containing the poison gas is both broken and not broken. The cat is both dead and alive!

This may sound absurd, but it's the logical extension of the particle being both decayed and not decayed. Schrödinger believed that quantum physics couldn't describe such an insane world, but that the cat had to be either completely alive or completely dead even before the box is opened and observed.

After you open the box, according to this interpretation, the cat's state becomes well defined one way or the other, but in the absence of a measurement, it's in both states. Though Schrödinger's cat experiment was created to oppose this interpretation of quantum mechanics, it has become the most dramatic example used to illustrate the strange quantum nature of reality.

Does Anyone Know What Quantum Theory Means?

Quantum physics is based on experimental evidence, much of which was obtained in the first half of the 20th century. The odd behavior has been seen in laboratories around the world, continually agreeing with the theory, despite all common sense. The really strange behavior occurs only on small scales; when you get to the size of cats, the quantum phenomena seems to always take on a definite value. Still, even today, the exact meaning of this strange quantum behavior is up in the air — something that doesn't trouble most modern physicists who work on these problems.

Some physicists hope that a "theory of everything," perhaps even string theory, may provide clear explanations for the underlying physical meaning of quantum physics. Among them, Lee Smolin has cited string theory's failure to explain quantum physics as a reason to look elsewhere for a fundamental theory of the universe — a view that is certainly not maintained by the majority of string theorists. Most string theorists believe that what matters is that quantum physics works (that is, it makes predictions that match experiment) and the philosophical concerns of why it works are less important. All of the interpretations of why quantum physics work yield the same experimental predictions, so they are effectively equivalent.

Einstein spent the last 30 years of his life railing against the scientific and philosophical implications of quantum physics. This was a lively time of debate in physics, as he and Niels Bohr sparred back and forth. "God does not play dice with the universe," Einstein was quoted as saying. Bohr replied, "Einstein, stop telling God what to do!"

A similar era may be upon us now, as theoretical physicists attempt to uncover the fundamental principles that guide string theory. Unlike quantum theory, there are few (if any) experimental results to base new work on, but there are many Einsteinian critics — again, on both scientific and philosophical grounds. (We get to them in Part V.)

Even with a firm theory that clearly works, physicists continue to question what quantum physics really means. What is the physical reality behind the mathematical equations? What actually happens to Schrödinger's cat? Some physicists hope that string theory may provide an answer to this question, though this is far from the dominant view. Still, any successful attempt to extend quantum physics into a new realm could provide unexpected insights that may resolve the questions.

Interactions transform quantum systems: The Copenhagen interpretation

The *Copenhagen interpretation* represents the orthodox view of quantum physics as it's taught in most undergraduate level courses, and it's mostly how I've interpreted quantum physics in this chapter: An observation or measurement causes the wavefunction to collapse from a general state of probabilities to a specific state.

The name comes from the Copenhagen Institute in (you guessed it) Copenhagen, Denmark, where Niels Bohr and his students helped form quantum physics in the 1920s and early 1930s, before World War II caused many to leave the Netherlands as they picked sides.

In today's talk, most physicists view the particles in the wavefunction as continually interacting with the world around them. These interactions are enough to cause the wavefunction to go through a process called *decoherence,* which basically makes the wavefunction collapse into a definite value. In other words, the very act of interacting with other matter causes a quantum system to become a classical system. Only by carefully isolating the quantum system to avoid such interactions will it remain in a coherent state, staying as a wave long enough to exhibit exotic quantum behaviors such as interference.

Under this explanation, you don't have to open the box for Schrödinger's cat to take on a definite state. The Geiger counter is probably where the breakdown occurs, and reality makes a "choice" of whether the particle has or has not decayed. Decoherence of the wavefunction takes place well before it ever reaches the cat.

If no one's there to see it, does the universe exist? The participatory anthropic principle

The *participatory anthropic principle (PAP)* was proposed by the physicist John Archibald Wheeler when he said that people exist in a "participatory universe." In Wheeler's (extremely controversial) view, an actual observer is needed to cause the collapse of the wavefunction, not just bits and pieces bouncing into each other.

This stance goes significantly further than the strict tenets of the Copenhagen interpretation, but it can't be completely dismissed when you look in depth at the quantum evidence. If you never look at the quantum system, then for all intents and purposes it always stays a quantum system. Schrödinger's cat really is both alive and dead until a person looks inside the box.

To John Barrow and Frank Tipler (in their popular and widely controversial 1986 book *The Anthropic Cosmological Principle*), this means that the universe itself comes into being only if someone is there to observe it. Essentially, the universe requires some form of life present for the wavefunction to collapse in the first place, meaning that the universe itself could not exist without life in it.

Most physicists believe that the PAP approach places humans in a crucial role in the universe, a stance which went out of favor when Copernicus realized Earth wasn't the center of the universe. As such, they (rightly, I believe) dismiss this interpretation in favor of those where humans aren't necessary components of the universe.

This is an especially strong statement of a concept known as the *anthropic principle*. Recent discoveries in string theory have caused some theoretical physicists who were once strongly opposed to any form of anthropic principle to begin to adopt weaker versions of the anthropic principle as the only means of making predictions from the vast array of string theory possibilities. I explain more about this concept in Chapter 11.

All possibilities take place: The many worlds interpretation

In contrast, the *many worlds interpretation (MWI)* of Hugh Everett III proposes that the wavefunction never actually collapses, but all possibilities become actualities — just in alternate realities. The universe is continually splitting apart as every quantum question is resolved in every possible way across an immense multiverse of parallel universes.

This is one of the most unusual concepts to come out of quantum physics, but it has its own merit. Like the work of Einstein described in Chapter 6, Everett arrived at this theory in part by taking the mathematics of quantum theory and assuming it could be taken literally. If the equation shows that there are two possibilities, then why not assume that there are two possibilities?

When you look inside the box, instead of something odd happening to the quantum system, you actually become part of the quantum system. You now exist in two states — one state that has found a dead cat and one state that has found a living cat.

Though these parallel universes sound like the stuff of science fiction, a related concept of parallel universes may arise as a prediction of string theory. In fact, it's possible that there are a vast number of parallel universes — a vast multiverse. More on this in Chapter 15.

What are the odds? Consistent histories

In the *consistent histories* view, the many worlds aren't actually realized, but the probability of them can be calculated. It eliminates the need for observers by assuming that the infinite complexity of the universe can't be fully dealt with, even mathematically, so it averages out over a large number of possible histories to arrive at the probabilities of the ones that are more probable, including the one universe that contains the outcome actually witnessed — our own.

Strictly speaking, the consistent history interpretation doesn't exclude the multiple worlds interpretation, but it only focuses on the one outcome you're sure of, rather than the infinite outcomes that you can only conjecture.

From a physical standpoint, this is similar to the idea of decoherence. The wavefunctions continually interact with particles just enough to keep all the possibilities from being realized. After you analyze all the possible paths, many of them cancel out, leaving only a couple of possible histories — the cat is either alive or dead. Making the measurement determines which one is the real history and which one was only a possibility.

Searching for more fundamental data: The hidden variables interpretation

One final interpretation is the *hidden variables interpretation,* where the equations of quantum theory are hiding another level of physical reality. The strange probabilities of quantum physics (under this explanation) are the result of our ignorance. If you understood this hidden layer, the system would be fully *deterministic.* (In other words, if you knew all the variables, you'd know exactly what was going to happen, and the quantum probabilities would go away.)

The first hidden variables theory was developed in the 1920s by Louis de Broglie, but a 1932 proof by John von Neumann showed that such theories couldn't exist in quantum physics. In 1952, physicist David Bohm used a mistake in this proof and reworked de Broglie's theory into his own variant (which has become the most popular version).

The core of Bohm's argument was a mathematical counterexample to the uncertainty principle, showing that quantum theory could be consistent with the existence of particles that had definite position and velocity. He assumed that these particles reproduced (on average) the results of the Schrödinger wavefunction. He was then able to construct a quantum potential wave that could guide the particles to behave in this way.

In Bohm's hidden variables theory, there is another hidden layer of physical law that is more fundamental than quantum mechanics. The quantum randomness would be eliminated if this additional layer could be understood. If such a hidden layer exists, it should, in principle, be possible for physics to someday reveal it in some way — perhaps through a "theory of everything." (Of course, the existence of either a "hidden layer" or "theory of everything" are ideas that aren't believed by most physicists today.)

Quantum Units of Nature — Planck Units

Physicists occasionally use a system of natural units, called *Planck units,* which are calculated based on fundamental constants of nature like Planck's constant, the gravitational constant, and the speed of light.

Planck's constant comes up often in discussing quantum physics. In fact, if you were to perform the mathematics of quantum physics, you'd find that little *h* variable all over the place. Physicists have even found that you can define a set of quantities in terms of Planck's constant and other fundamental constants, such as the speed of light, the gravitational constant, and the charge of an electron.

These Planck units come in a variety of forms. There is a Planck charge and a Planck temperature, and you can use various Planck units to derive other units such as the Planck momentum, Planck pressure, and Planck force . . . well, you get the idea.

For the purposes of the discussion of string theory, only a few Planck units are relevant. They are created by combining the gravitational constant, the speed of light, and Planck's constant, which makes them the natural units to use when talking about phenomena that involve those three constants, such as quantum gravity. The exact values aren't important, but here are the general scales of the relevant Planck units:

- Planck length: 10^{-35} meters (if a hydrogen atom were as big as our galaxy, the Planck length would be the size of a human hair)

- Planck time: 10^{-43} seconds (the time light takes to travel the Planck length — a very, *very* short period of time)

- Planck mass: 10^{-8} kilograms (about the same as a large bacteria or very small insect)

- Planck energy: 10^{28} electronvolts (roughly equivalent to a ton of TNT explosive)

Keep in mind that the exponents represent the number of zeroes, so the Planck energy is a 1 followed by 28 zeroes, in electronvolts. The most powerful particle accelerator on Earth, the Large Hadron Collider that came online briefly in 2008 can produce energy only in the realm of TeV — that is, a 1 followed by 12 zeroes, in electronvolts.

The negative exponents, in turn, represent the number of decimal places in very small numbers, so the Planck time has 42 zeroes between the decimal point and the first non-zero digit. It's a very small amount of time!

Some of these units were first proposed in 1899 by Max Planck himself, before either relativity or quantum physics. Such proposals for *natural units* — units based on fundamental constants of nature — had been made at least as far back as 1881. Planck's constant makes its first appearance in the physicist's 1899 paper. The constant would later show up in his paper on the quantum solution to the ultraviolet catastrophe.

Planck units can be calculated in relation to each other. For example, it takes exactly the Planck time for light to travel the Planck length. The Planck energy is calculated by taking the Planck mass and applying Einstein's $E = mc^2$ (meaning that the Planck mass and Planck energy are basically two ways of writing the same value).

In quantum physics and cosmology, these Planck units sneak up all the time. Planck mass represents the amount of mass needed to be crammed into the Planck length to create a black hole. A field in quantum gravity theory would be expected to have a vacuum energy with a density roughly equal to one Planck energy per cubic Planck length — in other words, it's 1 Planck unit of energy density.

Why are these quantities so important to string theory?

The Planck length represents the distance where the smoothness of relativity's space-time and the quantum nature of reality begin to rub up against each other. This is the quantum foam I explain in Chapter 2. It's the distance where the two theories each, in their own way, fall apart. Gravity explodes to become incredibly powerful, while quantum fluctuations and vacuum energy run rampant. This is the realm where a theory of quantum gravity, such as string theory, is needed to explain what's going on.

Planck units and Zeno's paradox

If the Planck length represents the shortest distance allowed in nature, it could be used to solve the ancient Greek puzzle called *Zeno's paradox*. Here is the paradox:

You want to cross a river, so you get in your boat. To reach the other side, you must cross half the river. Then you must cross half of what's left. Now cross half of what's left. No matter how close you get to the other side of the river, you will always have to cover half that distance, so it will take you forever to get across the river, because you have to cross an infinite number of halves.

The traditional way to solve this problem is with calculus, where you can show that even though there are an infinite number of halves, it's possible to cross them all in a finite amount of time. (Unfortunately for generations of stymied philosophers, calculus was invented by Newton and Leibnitz 2,000 years after Zeno posed his problem.)

As it turns out, during my sophomore year I solved Zeno's paradox in my calculus course the same semester that I learned about Planck units in my modern physics course. It occurred to me that if the Planck length were really the shortest distance allowed by nature, the quantum of distance, it offered a physical resolution to the paradox.

In my view, when your distance from the opposite shore reaches the Planck length, you *can't* go half anymore. Your only options are to go the whole Planck length or go nowhere. In essence, I pictured you "slipping" along that last tiny little bit of space without ever actually cutting the distance in half.

When I first came up with this idea as an undergraduate physics major, I was extremely impressed with myself. I have since learned that I'm not the only person to have come up with this connection between Planck length and Zeno's paradox. Despite that, I'm still somewhat impressed with myself.

In some sense, these units are sometimes considered to be quantum quantities of time and space, and perhaps some of the other quantities as well. Mass and energy clearly come in smaller scales, but time and distance don't seem to get much smaller than the Planck time and Planck length. Quantum fluctuations, due to the uncertainty principle, become so great that it becomes meaningless to even talk about something smaller. (See the nearby sidebar "Planck units and Zeno's paradox.")

In most string theories, the length of the strings (or length of compactified extra space dimensions) are calculated to be roughly the size of the Planck length. The problem with this is that the Planck length and the Planck energy are connected through the uncertainty principle, which means that to explore the Planck length — the possible length of a string in string theory — with precision, you'd introduce an uncertainty in energy equal to the Planck energy.

This is an energy 16 orders of magnitude (add 16 zeroes!) more powerful than the newest, most powerful particle accelerator on Earth can reach. Exploring such small distances requires a vast amount of energy, far more energy than we can produce with present technology.

Chapter 8

The Standard Model of Particle Physics

During the mid-1900s, physicists further explored the foundations of quantum physics and the components of matter. They focused on the study of particles in a field that became known as *particle physics*. More of these itty-bitty particles seemed to spring up every time physicists looked for them! By 1974, physicists had determined a set of rules and principles called the *Standard Model of particle physics* — a model that includes all interactions except for gravity.

Here I explore the Standard Model of particle physics and how it relates to string theory. Any complete string theory will have to include the features of the Standard Model and also extend beyond it to include gravity as well. In this chapter, I describe the structure of the atom, including the smaller particles contained within it, and the scientific methods used to explain the interactions holding matter together. I identify the two categories of particles that exist in our universe, fermions and bosons, and the different rules they follow. Finally, I point out the problems that remain from the Standard Model, which string theory hopes to resolve.

The topics related to the development of the Standard Model of particle physics are detailed and fascinating in their own right, but this book is about string theory. So my review of the material in this chapter is necessarily brief and is in no way intended to be a complete look at the subject. Many of the initial topics regarding the discovery of the structure of the atom are recounted in *Einstein For Dummies* (Wiley), and many other popular books are available to explore some of the more involved concepts of particle physics that come along later.

Atoms, Atoms, Everywhere Atoms: Introducing Atomic Theory

Physicist Richard P. Feynman once said that if he could boil down the most important principles of physics to a single sentence, it would be, "All things are made of atoms." (He actually goes on to expand on this, meaning that he actually boiled physics down to a compound sentence. For our immediate purposes, this first bit is enough.) The structure of atoms determines fundamental properties of matter in our universe, such as how atoms interact with each other in chemical combinations. The study of physics at the scale of an atom is called *atomic theory,* or atomic physics. Though this is several scales above the scale that string theory operates on, understanding the smaller structure of matter requires some level of understanding of the atomic-level structure.

Ancient Greeks considered the question of whether you could divide an object forever. Some — such as the fifth century B.C. philosopher Democritus — believed that you would eventually reach a smallest chunk of matter that couldn't be divided any more, and they called these smallest chunks *atoms.*

Aristotle's view that matter was composed of five basic elements was adopted by most philosophers of the time and remained the dominant way of thinking for many years, well into the time that "natural philosophy" began its transition into "science." After all, no scientists or philosophers had ever seen a smallest chunk of matter, so there really wasn't any reason to suppose they existed.

This began to change in 1738 when Swiss mathematician David Bernoulli explained how pressurized gas behaved by assuming that gas was made up of tiny particles. The heat of a gas was related to the speed of the particles. (This built on the work of Robert Boyle, nearly a century earlier.)

In 1808, British chemist John Dalton tried to explain the behavior of *elements* — substances that can't be chemically broken down into simpler substances — by assuming that they were made up of atoms.

According to Dalton, each atom of an element was identical to other atoms of the same element, and they combined together in specific ways to form the more complex substances we see in our universe.

Over the next century, evidence for the atomic theory mounted (see the sidebar "Einstein's contribution to atomic theory"). The complex structures formed by different atoms were called *molecules,* though the exact mechanism for how atoms formed molecules was still unclear.

Einstein's contribution to atomic theory

As if he weren't credited with enough, Albert Einstein is also frequently cited as the person who provided some of the last definitive support for the atomic theory of matter in two of his 1905 papers.

One of the papers was his PhD thesis, in which he calculated the approximate mass of an atom and the size of sugar molecules. This work earned him his doctorate from the University of Zurich.

The other paper involved analyzing random motion in smoke and liquids. This type of motion is called *Brownian motion* and had puzzled physicists for some time. Einstein pictured the motion as the result of atoms of smoke or liquid being jostled around by atoms of the surrounding gas or liquid, which explained the phenomenon perfectly. His predictions were supported by experimental findings.

It took more than 150 years from the time of Bernoulli for physicists to fully adopt the atomic model. Then, as you find out in the next section, after it was finally adopted, it was found to be incomplete! The complications arising in the study of string theory may well prove to take just as long, and perhaps ultimately be just as incomplete. But that doesn't mean they're necessarily "wrong," any more than atomic theory is "wrong."

Popping Open the Atomic Hood and Seeing What's Inside

Today scientists know that these atoms are not, as the Greeks imagined, the smallest chunks of matter. Scientists quickly realized that atoms had multiple parts inside of them:

- ✔ Negatively charged electrons circling the nucleus
- ✔ Positively charged nucleus

The particles that compose the nucleus (it's made up of smaller pieces, too) and electrons are among the particles, along with several others, that the Standard Model of particle physics explains, and ultimately that string theory should also explain.

Discovering the electron

The *electron* is a negatively charged particle contained within the atom. It was discovered in 1897 by British physicist J.J. Thomson, though charged particles (including the name "electron") had been hypothesized earlier.

Some physicists had already hypothesized that units of charge might be flowing around in electrical apparatus. (Benjamin Franklin proposed such an idea as early as the 1700s.) Technology only caught up to this idea in the late 1800s, with the creation of the cathode ray tube, shown in Figure 8-1.

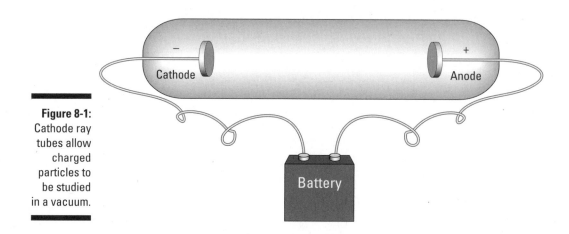

Figure 8-1:
Cathode ray tubes allow charged particles to be studied in a vacuum.

In a cathode ray tube, a pair of metal disks is connected to a battery. The metal disks are placed inside a sealed glass tube that contains no air — a vacuum tube. The electrical voltage causes one of the metal plates to become positively charged (an *anode*) and one to become negatively charged (the *cathode,* from which the device gets its name). Cathode ray tubes are the basis of traditional television and computer monitor tubes.

When the electrical current was switched on, the tube would begin to glow green. In 1897, Thomson was head of the Cavendish laboratory in Cambridge, England, and set about to test the properties of this cathode ray tube glow. He discovered that the glow was due to a beam of negatively charged particles flying between the plates. These negatively charged particles later came to be called electrons. Thomson also figured out that the electrons were incredibly light — 2,000 times lighter than a hydrogen atom.

Thomson not only discovered the electron, but he theorized that the electron was part of the atom (atoms weren't a completely accepted idea at the time) that somehow got knocked free from the cathode and flowed through the vacuum to the anode. With this discovery, scientists began discovering ways to explore the inside of atoms.

The nucleus is the thing in the middle

In the center of the atom is a dense ball of matter, called a *nucleus,* with a positive electrical charge. Shortly after electrons were discovered, it became clear that if you extracted an electron from an atom, the atom was left with a slightly positive electrical charge. For a while, the assumption was that the atom was a positively charged mass that contained negative electrons inside of it, like pieces of negatively charged fruit in a positively charged fruitcake. The entire fruitcake would be neutral unless you extracted some fruit from it. (Scientists of the day, being of a different dietary constitution than most of us today, explained it as plum pudding instead of fruitcake. Plum pudding or fruitcake — it unappetizingly amounts to roughly the same picture.)

In 1909, however, an experiment by Hans Geiger and Ernest Marsden, working under Ernest Rutherford, challenged this picture. These scientists fired positively charged particles at a thin sheet of gold foil. Most of the particles passed straight through the foil, but every once in a while one of them bounced back sharply. Rutherford concluded that the positive charge of the gold atom wasn't spread throughout the atom in the fruitcake model, but was concentrated in a small positively charged nucleus, and that the rest of the atom was empty space. The particles that bounced were the ones that hit this nucleus.

Watching the dance inside an atom

In trying to figure out the atom's structure, a natural model for scientists to look to was the planetary model, as shown in Figure 8-2. The electrons move around the nucleus in orbits. Physicist Niels Bohr determined that these orbits were governed by the same quantum rules that Max Planck had originally applied in 1900 — that energy had to be transferred in discrete packets.

In astronomy, the Earth and sun are attracted to each other by gravity, but because Earth is in motion around the sun, they never come into contact. A similar model could explain why the negative and positive portions of the atom never came into contact.

The first planetary model was proposed in 1904 by Nobel Prize-winner Hantaro Nagaoka. It was based on the rings of Saturn and called the Saturnian model. Certain details of the model were disproved by experiment, and Nagaoka abandoned the model in 1908, but Ernest Rutherford revised the concept to create his own planetary model in 1911, which was more consistent with experimental evidence.

Figure 8-2:
The Rutherford-Bohr model of the atom has electrons moving in orbits around a positively charged nucleus.

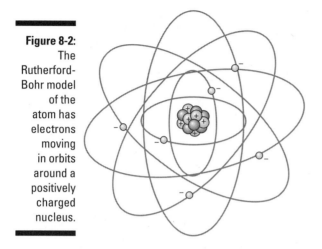

When atoms emitted electrons, the electron's energy followed certain precise patterns. Bohr realized in 1913 that this meant Rutherford's model required some revision. To fit the patterns, he applied the idea that energy was *quantized,* or bundled together in certain quantities, which allowed for stable orbits (instead of the collapsing orbits predicted by electromagnetism). Each electron could only exist in a certain, precisely defined energy state within its orbit. To go from one orbit to a different orbit required the electron to have enough energy to jump from one energy state to another.

Because of the quantum nature of the system, adding half the amount of energy to go from one orbit to another didn't move the electron halfway between those orbits. The electron remained in the first orbit until it received enough energy to kick it all the way into the higher-energy state. This is yet more of the strange behavior you've (hopefully) come to expect from quantum physics.

The Rutherford-Bohr model works pretty well in describing the hydrogen atom, but as atoms get more complex, the model begins to break down. Still, the basic principles hold for all atoms:

✔ A nucleus is at the center of an atom.

✔ Electrons move in orbits around the nucleus.

✔ The electron orbits are quantized (they have discrete energy levels) and are governed by the rules of quantum physics (though it would take several years for those rules to become developed, as described in Chapter 7).

The Quantum Picture of the Photon: Quantum Electrodynamics

The development of the theory of *quantum electrodynamics (QED)* was one of the great intellectual achievements of the 20th century. Physicists were able to redefine electromagnetism by using the new rules of quantum mechanics, unifying quantum theory and electromagnetic theory. Quantum electrodynamics was one of the first quantum approaches to a quantum field theory (described in the next section), so it introduced many features possessed by string theory (which is also a quantum field theory).

Quantum electrodynamics began with the attempt to describe particles in terms of quantum fields, starting in the late 1920s. In the 1940s, QED was completed three distinct times — by the Japanese physicist Sin-Itiro Tomonaga during World War II and later by American physicists Richard Feynman and Julian Schwinger. These three physicists shared the 1965 Nobel Prize in Physics for this work.

Dr. Feynman's doodles explain how particles exchange information

Though the principles of quantum electrodynamics were worked out by three individuals, the most famous founder of QED was undeniably Richard P. Feynman. Feynman was equally good at the mathematics and explanation of a theory, which resulted in his creation of *Feynman diagrams* — a visual representation of the mathematics that went on in QED.

Richard Phillips Feynman is one of the most interesting characters in 20th century physics, easily ranking with Einstein in personality, if not in pure fame. Early on in his career, Feynman made the conscious decision to only work on problems that he found interesting, something that certainly served him well. Fortunately for the world of physics, one of these problems was quantum electrodynamics.

Because electromagnetism is a field theory, the result of QED was a *quantum field theory* — a quantum theory that contains a value at every point in space. You can imagine that the mathematics of such a theory was intimidating, to say the least, even to those trained in physics and mathematics.

Feynman was brilliant not only with physical theory and mathematics, but also with explanation. One way he simplified things was through the application of his Feynman diagrams. Though the math was still complex, the diagrams meant you could begin talking about the physics without needing all the complexity of the equations. And when you did need the actual numbers, the diagrams helped organize your computations.

In Figure 8-3, you can see a Feynman diagram of two electrons approaching each other. The Feynman diagram is set on a Minkowski space, as introduced in Chapter 6, which depicts events in space-time. The electrons are the solid lines (called *propagators*), and as they get near to each other, a photon (the squiggly propagator; see Chapter 7 for the basics of photons) is exchanged between the two electrons.

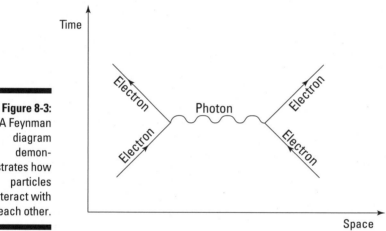

Figure 8-3:
A Feynman diagram demonstrates how particles interact with each other.

In other words, in QED two particles communicate their electromagnetic information by emitting and absorbing a photon. A photon that acts in this manner is called a *virtual photon* or a *messenger photon,* because it's created solely for the purpose of exchanging this information. This was the key insight of QED, because without this exchange of a photon, there was no way to explain how the information was communicated between the two electrons.

Also (and perhaps more important from a physics standpoint), a quantum field theory (at least those that seem to match our real world) quickly reaches infinity if distances become too small. To see how these infinities can arise, consider both the fact that electromagnetic forces get larger at small distances (infinitely larger at infinitely small distances) and also the distance and momentum relationship from the uncertainty principle of quantum mechanics (see Chapter 6 for details of the uncertainty principle). Even talking about the instances where two electrons are incredibly close to each other (such as within a Planck length) becomes effectively impossible in a world governed by quantum physics.

By quantizing electromagnetics, as QED does, Feynman, Schwinger, and Tomonaga were able to use the theory despite these infinities. The infinities were still present, but because the virtual photon meant that the electrons didn't need to get so close to each other, there weren't as many infinities, and the ones that were left didn't enter physical predictions. Feynman, Schwinger, and Tomonaga took an infinite theory and extracted finite predictions. One of the major motivations for the drive to develop a successful string theory is to go even further and get an actually finite theory.

The mathematical process of removing infinities is called *renormalization.* This is a set of mathematical techniques that can be applied to provide a very carefully defined limit for the continuum of values contained in the field. Instead of adding up all the infinite terms in the calculation and getting an infinite result, physicists have found that applying renormalization allows them to redefine parameters within the summation so it adds up to a finite amount! Without introducing renormalization, the values become infinite, and we certainly don't observe these infinities in nature. With renormalization, however, physicists get unambiguous predictions that are among the most precise and best-tested results in all of science.

Discovering that other kind of matter: Antimatter

Along with the understanding of quantum electrodynamics, there came a growing understanding that there existed *antimatter,* a different form of matter that was identical to known matter, but with opposite charge. Quantum field theory indicated that for each particle, there existed an antiparticle. The antiparticle of the electron is called the *positron.*

In 1928, physicist Paul Dirac was creating the quantum theory of the electron (a necessary precursor to a complete QED theory), when he realized that the equation only worked if you allowed these extra particles — identical to electrons but with opposite charge — to exist. Just four years later, the first positrons were discovered and named by Carl D. Anderson while he was analyzing cosmic rays.

The mathematics of the theory implied a symmetry between the known particles and identical particles with opposite charge, a prediction that eventually proved to be correct. The theory demanded that antimatter exist. String theory implies another type of symmetry, called supersymmetry (see Chapter 10), which has yet to be proved, but which many physicists believe will eventually be discovered in nature.

When antimatter comes in contact with ordinary matter, the two types of matter annihilate each other in a burst of energy in the form of a photon. This can also be depicted in QED with a Feynman diagram, as shown on the left side of Figure 8-4. In this view, the positron is like an electron that moves backward through time (as indicated by the direction of the arrow on the propagator).

Figure 8-4:
(Left) A particle and antiparticle annihilate each other, releasing a photon. (Right) A photon splits into a particle and antiparticle, which immediately annihilate each other.

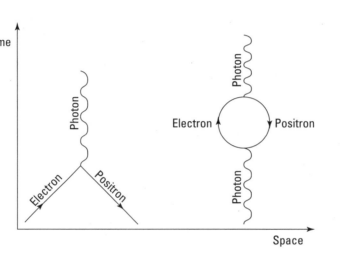

Sometimes a particle is only virtual

In quantum electrodynamics, *virtual particles* can exist briefly, arising from the energy fluctuations of the quantum fields that exist at every point in space. Some virtual particles — such as the photon in Figure 8-3 — exist just long enough to communicate information about a force. Other virtual particles spring into existence, seemingly for no purpose other than to make the lives of physicists more interesting.

The existence of virtual particles is one of the strangest aspects of physics, but it's a direct consequence of quantum physics. Virtual particles can exist because the uncertainty principle, in essence, allows them to carry a large fluctuation of energy, so long as they exist for only a brief period of time.

The right side of Figure 8-4 shows a pair of virtual particles — this time an electron and a positron. In some cases, a photon can actually split into an electron and positron and then recombine back into a photon.

The problem is that even though these particles are virtual, their effects have to be taken into account when performing calculations about what takes place in a given area. So no matter what you're doing, an infinite number of strange virtual particles are springing into and out of existence all around you, wreaking havoc with the smooth orderly calculations that you'd like to perform! (If this sounds familiar, it's because this is the quantum foam discussed in Chapter 2.)

Digging into the Nucleus: Quantum Chromodynamics

As quantum physics attempted to expand into the nucleus of the atom, new tactics were required. The quantum theory of the atomic nucleus, and the particles that make it up, is called *quantum chromodynamics (QCD).* String theory arose out of an attempt to explain this same behavior.

In the explanation of QED in the previous section, the only participants in QED were the photon and the electron (and, briefly, the positron). In fact, QED attempted to simplify the situation by only analyzing these two aspects of the atom, which it could do by treating the nucleus as a giant, very distant object. With QED finally in place, physicists were ready to take a good hard look at the nucleus of the atom.

The pieces that make up the nucleus: Nucleons

The nucleus of an atom is composed of particles called *nucleons,* which come in two types: positively charged *protons* and the noncharged *neutrons.* The protons were discovered in 1919, while the neutrons were discovered in 1932.

The proton is about 1,836 times as massive as the electron. The neutron is about the same size as the proton, so the pair of them is substantially larger than the electron. Despite this difference in size, the proton and electron have identical electrical charges, but of opposite sign; the proton is positive while the electron is negative.

The growth of technology allowed for the design and construction of larger and more powerful *particle accelerators,* which physicists use to smash particles into each other and see what comes out. With great delight, physicists began smashing protons into each other, in the hopes of finding out what was inside of them.

In fact, this work on trying to uncover the secrets of these nucleons would lead directly to the first insights into string theory. A young physicist at CERN applied an obscure mathematical formula to describe the behavior of particles in a particle accelerator, and this is seen by many as the starting point of string theory. (These events are covered in more detail in Chapter 10.)

The pieces that make up the nucleon's pieces: Quarks

Today, the nucleons are known to be types of *hadrons,* which are particles made up of even smaller particles called *quarks.* The concept of quarks was independently proposed by Murray Gell-Mann and George Zweig in 1964 (though the name, taken from James Joyce's *Finnegan's Wake,* is pure Gell-Mann), which in part earned Gell-Mann the 1969 Nobel Prize in Physics. The quarks are held together by still other particles, called *gluons.*

In this model, both the proton and the neutron are composed of three quarks. These quarks have quantum properties, such as mass, electrical charge, and spin (see the next section for an explanation of spin). There are actually a total of six *flavors* (or types) of quarks, all of which have been experimentally observed:

- Up quark
- Down quark
- Charm quark
- Strange quark
- Top quark
- Bottom quark

The properties of the proton and neutron are determined by the specific combination of quarks that compose them. For example, a proton's charge is reached by adding up the electrical charge of the three quarks inside it — two up quarks and one down quark. In fact, every proton is made of two up quarks and one down quark, so they're all exactly alike. Every neutron is identical to every other neutron (composed of one up quark and two down quarks).

In addition to standard quantum mechanical properties (charge, mass, and spin), quarks have another property, which came out of the theory, called *color charge.* This is somewhat similar to electrical charge in principle, but it's an entirely distinct property of quarks. It comes in three varieties, named *red, green,* and *blue.* (Quarks don't actually have these colors, because they're much, much smaller than the wavelength of visible light. These are just names to keep track of the types of charge.)

Because QED describes the quantum theory of the electrical charge, QCD describes the quantum theory of the color charge. The color charge is the source of the name quantum chromodynamics, because "chroma" is Greek for "color."

In addition to the quarks, there exist particles called *gluons.* The gluons bind the quarks together, kind of like rubber bands (in a very metaphoric sense). These gluons are the gauge bosons for the strong nuclear force, just as the photons are the gauge bosons for electromagnetism (see the later section on gauge bosons for more on these particles).

Looking into the Types of Particles

Physicists have found a large number of particles, and one thing that proves useful is that they can be broken down into categories based on their properties. Physicists have found a lot of ways to do this, but in the following sections I briefly discuss some of the most relevant categories to string theory.

According to quantum mechanics, particles have a property known as *spin.* This isn't an actual motion of the particle, but in a quantum mechanical sense, it means that the particle always interacts with other particles as if it's rotating in a certain way. In quantum physics, spin has a numerical value that can be either an integer (0, 1, 2, and so on) or half-integer (½, ⅔, and so on). Particles that have an integer spin are called *bosons,* while particles that have half-integer spin are called *fermions.*

Particles of force: Bosons

Bosons, named after Satyendra Nath Bose, are particles that have an integer value of quantum spin. The bosons that are known act as carriers of forces in quantum field theory, as the photon does in Figure 8-3. The Standard Model of particle physics predicts five fundamental bosons, four of which have been observed:

- Photon
- Gluon (there are eight types of gluons)

✔ Z boson

✔ W boson (actually two particles — the W^+ and W^- bosons)

✔ Higgs boson (this one hasn't been found yet)

In addition, many physicists believe that there probably exists a boson called the *graviton,* which is related to gravity. The relationship of these bosons to the forces of physics are covered in the "Gauge Bosons: Particles Holding Other Particles Together" section later in this chapter.

Composite bosons can also exist; these are formed by combining together an even number of different fermions. For example, a carbon-12 atom contains six protons and six neutrons, all of which are fermions. The nucleus of a carbon-12 atom is, therefore, a composite boson. *Mesons,* on the other hand, are particles made up of exactly two quarks, so they are also composite bosons.

Particles of matter: Fermions

Fermions, named after Enrico Fermi, are particles that have a half-integer value of quantum spin. Unlike bosons, they obey the *Pauli exclusion principle,* which means that multiple fermions can't exist in the same quantum state.

While bosons are seen as mediating the forces of nature, fermions are particles that are a bit more "solid" and are what we tend to think of matter particles. Quarks are fermions.

In addition to quarks, there is a second family of fermions called *leptons.* Leptons are elementary particles that can't (so far as scientists know) be broken down into smaller particles. The electron is a lepton, but the Standard Model of particle physics tells us that there are actually three generations of particles, each heavier than the last. (The three generations of particles were predicted by theoretical considerations before they were discovered by experiment, an excellent example of how theory can precede experiment in quantum field theory.)

Also within each generation of particles are two flavors of quarks. Table 8-1 shows the 12 types of fundamental fermions, all of which have been observed. The numbers shown in Table 8-1 are the masses, in terms of energy, for each of the known particles. (Neutrinos have virtually, but not exactly, zero mass.)

Table 8-1	Elementary Particle Families for Fermions			
	Quarks		*Leptons*	
First Generation	Up Quark	Down Quark	Electron Neutrino	Electron
	3 MeV	7 MeV		0.5 MeV
Second Generation	Charm Quark	Strange Quark	Muon Neutrino	Muon
	1.2 GeV	120 MeV		106 MeV
Third Generation	Top Quark	Bottom Quark	Tau Neutrino	Tau
	174 GeV	4.3 GeV		1.8 GeV

There are also, of course, composite fermions, made when an odd number of fermions combine to create a new particle, such as how protons and neutrons are formed by combining quarks.

Gauge Bosons: Particles Holding Other Particles Together

In the Standard Model of particle physics, the forces can be explained in terms of gauge theories, which possess certain mathematical properties. These forces transmit their influence through particles called *gauge bosons*. String theory allows gravity to be expressed in terms of a gauge theory, which is one of its benefits. (One example of this is the AdS/CFT correspondence discussed in Chapter 11.)

Throughout the development of the Standard Model, it became clear that all the forces (or, as many physicists prefer, *interactions*) in physics could be broken down into four basic types:

- Electromagnetism
- Gravity
- Weak nuclear force
- Strong nuclear force

The electromagnetic force and weak nuclear force were consolidated in the 1960s by Sheldon Lee Glashow, Abdus Salam, and Steven Weinberg into a single force called the *electroweak force.* This force, in combination with quantum chromodynamics (which defined the strong nuclear force), is what physicists mean when they talk about the Standard Model of particle physics.

One key element of the Standard Model of particle physics is that it's a *gauge theory*, which means certain types of symmetries are inherent in the theory; in other words, the dynamics of the system stay the same under certain types of transformations. A force that operates through a gauge field is transmitted with a *gauge boson*. The following gauge bosons have been observed by scientists for three of the forces of nature:

- ✔ Electromagnetism — photon
- ✔ Strong nuclear force — gluon
- ✔ Weak nuclear force — Z, W^+, and W^- bosons

In addition, gravity can be written as a gauge theory, which means that there should exist a gauge boson that mediates gravity. The name for this theoretical gauge boson is the *graviton*. (In Chapter 10, you see how the discovery of the graviton in the equations of string theory led to its development as a theory of quantum gravity.)

Exploring the Theory of Where Mass Comes From

In the Standard Model of particle physics, particles get their mass through something called the *Higgs mechanism*. The Higgs mechanism is based on the existence of a *Higgs field*, which permeates all of space. The Higgs field creates a type of particle called a *Higgs boson*. For the Higgs field to create a Higgs boson takes a lot of energy, and physicists have so far been unable to create one — so it's the only particle predicted by the Standard Model of particle physics that hasn't been observed. This, together with attempts to find new particles, such as those motivated by string theory, are among the major reasons why scientists need advanced particle accelerators for more high-energy experiments.

The weak nuclear force falls off very rapidly above short distances. According to quantum field theory, this means that the particles mediating the force — the W and Z bosons — must have a mass (as opposed to the gluons and photons, which are massless).

The problem is that the gauge theories described in the preceding section are mediated only by massless particles. If the gauge bosons have mass, then a gauge theory can't be sensibly defined. The Higgs mechanism avoids this problem by introducing a new field called the Higgs field. At high energies, where the gauge theory is defined, the gauge bosons are massless, and the theory works as anticipated. At low energies, the field triggers broken symmetries that allow the particles to have mass.

If the Higgs field does exist, it would create particles known as Higgs bosons. The mass of the Higgs boson isn't something that the theory tells us, but most physicists anticipate it to be found in the range of 150 GeV. Fortunately, this is within the realm of what we can experimentally search for. Finding the Higgs boson would be the final confirmation of the Standard Model of particle physics.

The Higgs mechanism, Higgs field, and Higgs boson are named after Scottish physicist Peter Higgs. Though he wasn't the first to propose these concepts, he's the one they were named after, which is just one of those things that sometimes happens in physics.

 For a discussion on the Higgs mechanism in depth, I recommend Lisa Randall's *Warped Passages: Unraveling the Mysteries of the Universe's Hidden Dimensions.* Chapter 10 of that book is devoted entirely to this topic. You could also look to *The God Particle: If the Answer is the Universe, What is the Question?* by Nobel Laureate Leon Lederman and Dick Teresi, which is devoted entirely to the topic of the search for the Higgs boson.

From Big to Small: The Hierarchy Problem in Physics

The Standard Model of particle physics is an astounding success, but it hasn't answered every question that physics hands to it. One of the major questions that remains is the *hierarchy problem,* which seeks an explanation for the diverse values that the Standard Model lets physicists work with. Many physicists feel that string theory will ultimately be successful at resolving the hierarchy problem.

For example, if you count the theoretical Higgs boson (and both types of W bosons), the Standard Model of particle physics has 18 elementary particles. The masses of these particles aren't predicted by the Standard Model. Physicists had to find these by experiment and plug them into the equations to get everything to work out right.

If you look back at Table 8-1, you notice three families of particles among the fermions, which seems like a lot of unnecessary duplication. If we already have an electron, why does nature need to have a muon that's 200 times as heavy? Why do we have so many types of quarks?

Beyond that, when you look at the energy scales associated with the quantum field theories of the Standard Model, as shown in Figure 8-5, even more questions may occur to you. Why is there a gap of 16 orders of magnitude (16 zeroes!) between the intensity of the Planck scale energy and the weak scale?

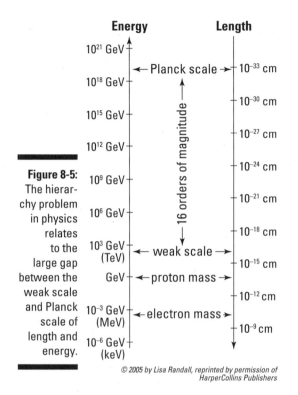

Figure 8-5:
The hierarchy problem in physics relates to the large gap between the weak scale and Planck scale of length and energy.

At the bottom of this scale is the *vacuum energy,* which is the energy generated by all the strange quantum behavior in empty space — virtual particles exploding into existence and quantum fields fluctuating wildly due to the uncertainty principle.

The hierarchy problem occurs because the fundamental parameters of the Standard Model don't reveal anything about these scales of energy. Just as physicists have to put the particles and their masses into the theory by hand, so too have they had to construct the energy scales by hand. Fundamental principles of physics don't tell scientists how to transition smoothly from talking about the weak scale to talking about the Planck scale.

As I explain in Chapter 2, trying to understand the "gap" between the weak scale and the Planck scale is one of the major motivating factors behind trying to search for a quantum gravity theory in general, and string theory in particular. Many physicists would like a single theory that could be applied at all scales, without the need for renormalization (the mathematical process of removing infinities), or at least to understand what properties of nature determine the rules that work for different scales. Others are perfectly happy with renormalization, which has been a major tool of physics for nearly 40 years and works in virtually every problem that physicists run into.

Chapter 9

Physics in Space: Considering Cosmology and Astrophysics

*O*ne of mankind's first scientific acts was probably to look into the heavens and ask questions about the nature of that expansive universe. Today, scientists are still fascinated by these questions, and with good reason. Though we know much more than our cave-dwelling ancestors did about what makes up the heavens, the black space between the stars still holds many mysteries — and string theory is at the heart of the search for the answers to many of these mysteries.

In this chapter, you find out what physicists, astronomers, astrophysicists, and cosmologists have uncovered about the workings of the universe independent of string theory. As these scientists have discovered how the universe works, their findings have led to more difficult questions, which string theorists hope to answer. I cover some of these more complex points about the universe in Chapter 14. This chapter gives you the background that will help you understand the ties between cosmology, astrophysics, and string theory.

In the following pages, I explore the consequences of Einstein's relativity, where scientists find that the universe seems to have had a beginning. At this point, scientists were able to determine where the particles in our universe come from. The theory of the universe's origin grows more complex with the introduction of a rapidly expanding early universe. I also introduce you to two of cosmology's biggest mysteries: the presence of unseen dark matter and of repulsive gravity in the form of dark energy. Finally, I provide a glimpse into black holes, objects that later become important to string theory.

Creating an Incorrect Model of the Universe

Before string theory, there was Einstein's relativity, and before that was Newton's gravity, and for about two centuries before Newton, the laws governing the universe were believed, by most of the Western world, to be those set out by Aristotle. Understanding the later revolutions in cosmology starts with the original models of the universe developed by the ancient Greeks.

Aristotle assigns realms to the universe

Aristotle pictured a universe that was made of a substance called the *aether* (see Chapter 5 for more about this elusive element). The heavens, to him and his followers, were a place of unsurpassed geometric elegance and beauty that didn't change over time.

In some ways, Aristotle is seen as one of the first scientists. He spent a great deal of time discussing the importance of observation to understand nature. Aristotle described the universe as containing five fundamental elements: earth, air, fire, water, and aether. The heavens were the realm of aether, but we were stuck down with the earth, air, fire, and water.

Aristotle knew Earth was a sphere, and he thought that each element had a natural location within that sphere, as shown in Figure 9-1. The natural location of the earth element was at the center of the sphere — this was considered the earth realm. Next came the water realm, followed by the air, fire, and finally aether realms. (The moon resided somewhere on the border of the aether realm, probably right on the edge of the fire realm.)

The clouds — composed of air and water elements — drifted in the air, along the border of the air and water realms. You can mix water and earth to make mud, but the earth part tends to eventually settle on the bottom because the earth realm is beneath the water realm. When a fire was ignited, the flames reached up into the sky in an effort to reach the fire realm, where the sun resided.

In Aristotle's model, the outermost sphere was the aether realm, relatively untouched by the mundane elements, aside from the moon (hardly mundane) and its border with the fire realm. It was a perfect realm, which contained the stars, fixed in place on a serene, eternal background. This belief defined the heavens for well over a thousand years.

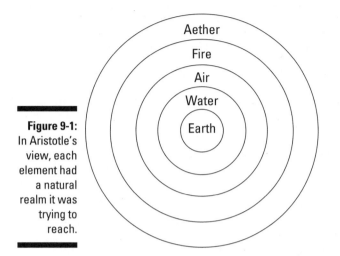

Figure 9-1:
In Aristotle's view, each element had a natural realm it was trying to reach.

Ptolemy puts Earth at the center of the universe (and the Catholic Church agrees)

The cosmological model of the stars' movements built on Aristotle's philosophy was called the *Ptolemaic model,* after the man who invented it.

Ptolemy lived in Roman Egypt during the second century AD, doing his principle work in the city of Alexandria. His book on astronomy, the *Almagest* (which roughly translates to "the greatest") was written in approximately 150 AD. The greatest achievement of this volume was to attempt to describe the motion of the heavens in precise mathematical language.

The model described by Ptolemy, and maintained by most scholars until the time of Copernicus and Galileo, was a *geosynchronous model* of the heavens, in which Earth was at the center of the universe. The reason for such a model is obvious from Aristotle's elemental spheres depicted in Figure 9-1: Earth has a distinct, unique place in the universe.

In Ptolemy's geosynchronous model, the moon, planets, and sun are mounted on rotating spheres around Earth. Beyond the planets is the largest sphere, which has the stars mounted on it. This model accurately predicted the motion of the planets, so it was well received.

The Catholic Church adopted this model of the universe for a number of reasons. One is that it provided a way for the sun to be "held still" in the sky to match a Biblical account. Another reason is that the theory said nothing about what was outside the star-laden spheres, so it left plenty of room for heaven and hell.

So many scientists, so many names

The names for different types of space scientists can get rather confusing. Gone are the days when anyone who looked through a telescope could be called an astronomer. The distinction between astronomer and astrophysicist is pretty much history, and the line between astrophysicist and cosmologist gets blurry in the realm of string theory. The term used is often chosen out of personal preference, but there are some guidelines:

✔ **Astronomer:** This is the classical term for a scientist who studies the heavens. Since Galileo, optical telescopes have been the primary tool used to examine celestial bodies. Today, the telescopes can be radio, x-ray, or gamma ray telescopes, which see light in the nonvisual spectrum. Traditionally, astronomers have devoted more time to classification and description of bodies in space than to attempting to explain the phenomena.

✔ **Astrologer:** In the time of Ptolemy through Copernicus, the terms *astrologer* and *astronomer* were essentially synonyms. Since Copernicus, they have become more distinct; today they represent radically different disciplines, with astrology well outside the bounds of science. An astrologer tries to find a connection between human behaviors and the motion of celestial bodies, generally with a vague or supernatural mechanism introduced as the basis for these connections. See *Astrology For Dummies,* 2nd Edition, by Rea Orion (Wiley) for more information on the field.

✔ **Astrophysicist:** This term applies to someone who studies the physics of interactions within and between stellar bodies. Astrophysicists seek to apply the principles of physics to create general laws governing the behavior of these interactions.

✔ **Cosmologist:** This term is used for a type of astrophysicist who focuses on the evolution of the universe — the processes of how the universe changes over time. A cosmologist rarely cares about a specific stellar body or solar system, and galaxies are frequently too parochial for these explorers of space. Cosmologists often focus their attention on theories that use unimaginably large scales of time, space, and energy. The study of the big bang, or the universe's end, is an example of the cosmologist's domain.

Perhaps most significantly, the Church embraced the belief that Earth and the heavens were made of different things. Our realm was special. In all of space there was nothing else quite like Earth, and certainly no other place that could give rise to anything resembling humanity. With the Catholic Church's official endorsement, the Ptolemaic model of the universe became not just a scientific theory, but a religious fact.

The Enlightened Universe: Some Changes Allowed

In the 1500s, the geosynchronous model was replaced with the *heliocentric model,* in which the sun was at the center of the solar system. (Heliocentric models had originally been proposed by Greeks such as Aristarchus, but Aristotle's model gained greater popularity.) The work of Nicholas Copernicus and Galileo Galilei was key to this revolution, which dislodged us from our special place at the center of the universe. The result has become known as the *Copernican principle,* which says that space looks the same no matter where you view it from.

Copernicus corrects what's where in the universe

The Ptolemaic model was based on the idea that all the celestial objects — planets, moons, stars, and so on — were on concentric spheres, each of which was centered on Earth. Over the centuries (from about 150 BC to 1500 AD), though, observations made it clear that this wasn't the case.

To preserve the Ptolemaic model, it was modified over the years. Celestial objects were mounted on spheres that were then mounted on other spheres. The very elegance that made the Ptolemaic model so appealing was gone, replaced with a mishmash of geometric nonsense that only partially conformed with scientific observations — which were growing more and more precise due to new technologies.

This was the prime time for a scientific revolution. The existing theory was failing, but without another system in place to adopt (the heliocentric models of Aristarchus were ignored, for some reason), the prevailing system continued to be modified in increasingly improbable ways (check out Chapter 4 for more on this process). In the case of the Ptolemaic model, the fact that contradicting it was heretical didn't help incite a scientific revolution either.

In his book, *On the Revolutions of the Celestial Spheres,* the Polish astronomer Nicholas Copernicus explained his heliocentric model, making it clear that the sun, not Earth, sat at center stage. He still used spheres, though, and made other assumptions that haven't born the test of time, but it was a major improvement over the Ptolemaic model.

Copernicus published his heliocentric model upon his death in 1543, fearing retribution from the Church if he published it earlier (although he did hand out versions of the theory to friends about 30 years earlier). Some Indian writers made this heliocentric claim as far back as the seventh century AD, and some Islamic astronomers and mathematicians studied this idea as well, but it's unclear to what degree Copernicus was aware of their work.

Copernicus was a theorist, not an observational astronomer. His key insight was the idea that Earth didn't have a distinct position within the universe, a concept that was named the *Copernican principle* in the mid-20th century.

Beholding the movements of heavenly bodies

One of the greatest observational astronomers of this revolutionary age was Tycho Brahe, a Danish nobleman who lived from 1546 to 1601. Brahe made an astounding number of detailed astronomical observations. He used his family's wealth to found an observatory that corrected nearly every astronomical record of the time, including those in Ptolemy's *Almagest*.

Using Brahe's measurements, his assistant Johannes Kepler was able to create rules governing the motion of the planets in our solar system. In his three laws of planetary motion, Kepler realized that the planetary orbits were elliptical rather than circular.

More importantly, Kepler discovered that the motion of the planets wasn't uniform. A planet's speed changes as it moves along its elliptical path. Kepler showed that the heavens were a dynamic system, a detail that later helped Newton show that the sun constantly influences the planets' motion.

Galileo, by using the telescope, later realized that other planets had moons and determined that the heavens weren't static. The Catholic Church charged him with heresy. To get away with only house arrest, Galileo was forced to recant his observations about the movements of heavenly bodies. Reportedly, his final words on his death bed were, "But they *do* move!" (Some versions of this story indicate that he uttered this statement upon being sentenced, so it may be a myth.)

Galileo's work, together with Kepler's, laid the foundation for Isaac Newton's law of gravity. With gravity introduced, the final nail had been placed in the scientific consensus behind the geosynchronous view. Astronomers and physicists now knew that Earth circled the sun, as the heliocentric model described. (The Catholic Church officially endorsed the heliocentric view in the 19th century. In 1992, Pope John Paul II officially apologized for Galileo's treatment.)

Introducing the Idea of an Expanding Universe

Even two centuries after Newton, Albert Einstein was strongly influenced by the concept of an unchanging universe. His general theory of relativity predicted a dynamic universe — one that changed substantially over time — so he introduced a term, called the *cosmological constant,* into the theory to make the universe static and eternal. This would prove to be a mistake when, several years later, astronomer Edwin Hubble discovered that the universe was expanding! Even today, the consequence of the cosmological constant in general relativity has enormous impact upon physics, causing string theorists to rethink their whole approach.

The equations of general relativity that Einstein developed showed that the very fabric of space was expanding or contracting. This made no sense to Einstein, so in 1917 he added the cosmological constant to the equations. This term represented a form of repulsive gravity that exactly balanced out the attractive pull of gravity.

When Hubble showed that the universe was indeed expanding, Einstein called the introduction of the cosmological constant his "biggest blunder" and removed it from the equations. This concept would return over the years, however, as you see in the "Dark Energy: Pushing the Universe Apart" section later in this chapter. With the discovery of dark energy, Einstein's "blunder" was found to be a necessary parameter in the theory (even though physicists for most of a century assumed the cosmological constant's value was zero).

Discovering that energy and pressure have gravity

In Newton's gravity, bodies with mass were attracted to each other. Einstein's relativity showed that mass and energy were related. Therefore, mass and energy both exerted gravitational influence. Not only that, but it was possible that space itself could exert a pressure that warped space. Several models were constructed to show how this energy and pressure affected the expansion and contraction of space.

When Einstein created his first model based on the general theory of relativity, he realized that it implied an expanding universe. At the time, no one had any particular reason to think the universe was expanding, and Einstein assumed that this was a flaw in his theory.

Einstein's general relativity equations allowed for the addition of an extra term while remaining mathematically viable. Einstein found that this term could represent a positive energy (or negative pressure) uniformly distributed throughout the fabric of space-time itself, which would act as an *antigravity,* or repulsive form of gravity. This term was chosen to precisely cancel out the contraction of the universe, so the universe would be static (or unchanging in time).

In 1917, the same year Einstein published his equations containing the cosmological constant, Dutch physicist Willem de Sitter applied them to a universe without matter. (As I explain in Chapter 4, this is a frequent step in scientific analysis — you strip a scientific theory of all the complications and consider it in the simplest cases.)

In this *de Sitter space,* the only thing that exists is the energy of the vacuum — the cosmological constant itself. Even in a universe containing no matter at all, this means that space will expand. A de Sitter space has a positive value for the cosmological constant, which can also be described as a positive curvature of space-time. A similar model with a negative cosmological constant (or a negative curvature, in which expansion is slowing) is called an *anti-de Sitter space.* (More on the curvature of space-time in a bit.)

In 1922, the Russian physicist Aleksandr Friedmann turned his hand to solving the elaborate equations of general relativity, but decided to do so in the most general case by applying the *cosmological principle* (which can be seen as a more general case of the Copernican principle), which consists of two assumptions:

✔ The universe looks the same in all directions (it's *isotropic*).

✔ The universe is uniform no matter where you go (it's *homogenous*).

With these assumptions, the equations become much simpler. Einstein's original model and de Sitter's model both ended up being special cases of this more general analysis. Friedmann was able to define the solution depending on just three parameters:

✔ Hubble's constant (the rate of expansion of the universe)

✔ Lambda (the cosmological constant)

✔ Omega (average matter density in the universe)

To this day, scientists are trying to determine these values as precisely as they can, but even without real values they can define three possible solutions. Each solution matches a certain "geometry" of space, which can be represented in a simplified way by the way space naturally curves in the universe, as shown in Figure 9-2.

✔ **Closed universe:** There is enough matter in the universe that gravity will eventually overcome the expansion of space. The geometry of such a universe is a positive curvature, such as the sphere in the leftmost image in Figure 9-2. (This matched Einstein's original model without a cosmological constant.)

✔ **Open universe:** There isn't enough matter to stop expansion, so the universe will continue to expand forever at the same rate. This space-time has a negative curvature, like the saddle shape shown in the middle image in Figure 9-2.

✔ **Flat universe:** The expansion of the universe and the density of matter perfectly balance out, so the universe's expansion slows down over time but never quite stops completely. This space has no overall curvature, as shown in the rightmost image of Figure 9-2. (Friedmann himself didn't discover this solution; it was found years later.)

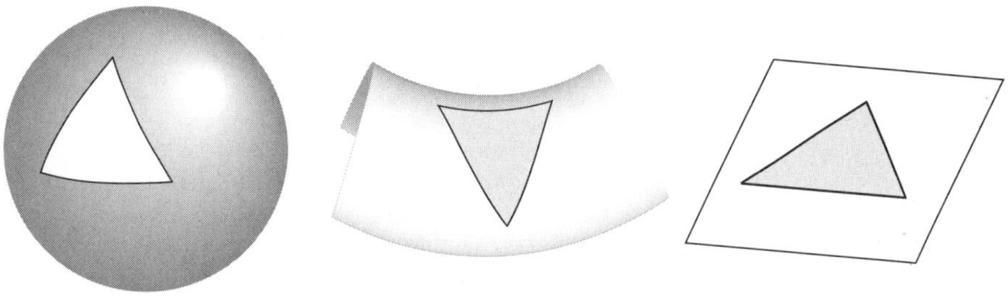

Figure 9-2: Three types of universes: closed, open, and flat.

These models are highly simplified, but they needed to be because Einstein's equations got very complex in cases where the universe was populated with a lot of matter, and supercomputers didn't yet exist to perform all the math (and even physicists want to go on dates every once in a while).

Hubble drives it home

In 1927, astronomer Edwin Hubble proved that the universe is expanding. With this new evidence, Einstein removed the cosmological constant from his equations.

Edwin Hubble had shown in 1925 that there were galaxies outside our own. Until that time, astronomers had observed white blobs of stars in the sky, which they called *nebulae,* but the astronomers disagreed about how far away they were. In his work at the Mount Wilson Observatory in California, Hubble proved that these were, in fact, distant galaxies.

While studying these distant galaxies, he noticed that the light from these distant stars had a wavelength that was shifted slightly toward the red end of the electromagnetic spectrum, compared to what he expected.

This is a consequence of the wave nature of light — an object that's moving (with respect to the observer) emits light with a slightly different wavelength. This is based on the *Doppler effect,* which is what happens to the wavelength of sound waves from a moving source. If you've ever heard a siren's pitch change as it approaches and passes you, you've experienced the Doppler effect.

In a similar way, when a light source is moving, the wavelength of the light changes. A *redshift* in light from a star means the star is moving away from the observer.

Hubble saw this redshift in the stars he observed, caused not only by the motion of the stars but by the expansion of space-time itself, and in 1929 determined that the amount of shift was related to the distance from Earth. The more distant stars were moving away faster than the nearby stars. Space itself was expanding.

Clearly, in this case, Einstein had been wrong and Friedmann had been right to explore all the possible scenarios predicted by general relativity. (Unfortunately, Friedmann died in 1925, so he never knew he was right.)

Finding a Beginning: The Big Bang Theory

It soon became evident that an expanding universe was once very much smaller — so small, in fact, that it was compressed down to a single point (or, at least, a very small area). The theory that the universe started from such a primordial point and has expanded ever since is known as the *big bang theory.* The theory was first proposed in 1927, but was controversial until 1965, when an accidental discovery supported the theory. Today, the most advanced astronomical observations show that the big bang theory is likely true. String theory will hopefully help physicists understand more precisely what happened in those early moments of the universe, so understanding the big bang theory is a key component of string theory's cosmological work.

The man originally responsible for the big bang theory was a Belgian priest and physicist, Georges Lemaître, who independently worked on theories similar to Friedmann's. Like Friedmann, Lemaître realized that the universe defined by general relativity would either expand or contract.

What's in a name?

The name "big bang" was given to the theory by Fred Hoyle, one of the theory's greatest critics. In a 1949 series of BBC radio broadcasts, Hoyle was speaking dismissively of the idea that everything in the universe was created in one sudden "big bang" in the distant past.

The name stuck, much to big bang theorists' dismay. Strictly speaking, the big bang theory doesn't include a bang. Rather, the theory states that a tiny primordial particle began to expand, creating the universe. There is neither big nor bang in this theory.

In 1927, Lemaître learned of Hubble's finding about distant galaxies moving away from Earth. He realized that this meant space was expanding, and he published a theory that came to be called the big bang theory. (See the nearby sidebar, "What's in a name?")

Because you know that space is expanding, you can run the video of the universe backward in time in your head (rewind it, so to speak). When you do this, you realize that the universe had to be much smaller than it is now. As the matter in the universe gets compressed into a smaller and smaller amount of space, the laws of thermodynamics (which govern the flow of heat) tell you that the matter had to be incredibly hot and dense.

The big bang theory reveals that the universe came from a state of dense, hot matter, but it tells nothing about how the matter got there, or whether anything else existed before the big bang (or even if the word "before" has any meaning when you're talking about the beginning of time). I explore these speculative topics in Chapters 14 and 15.

Bucking the big bang: The steady state theory

In opposition to the big bang theory, Fred Hoyle proposed an alternative theory, called the *steady state theory*. In this theory, new particles were continually being created. As space expanded, these new particles were created fast enough that the overall mass density of the universe remained constant.

To understand the reason for such a theory, you have to realize that few physicists thought it likely that a dense ball of matter could spring into existence out of nowhere, violating the law of conservation of mass (or conservation of mass-energy). That matter had to come from somewhere.

In Hoyle's view, if matter could be created out of nothing one time, then why not have it happen all the time?

Though Hoyle's steady state theory would ultimately fail, in trying to prove it Hoyle would prove himself worthy in the eyes of history by developing a theory about where the dense atoms of our universe come from (which I cover in the later section "Understanding where the chemical elements came from").

Going to bat for the big bang: Cosmic microwave background radiation

One of the major converts to the big bang theory was physicist George Gamow, who realized that if the theory were true, a residual trace of cosmic microwave background radiation (CMBR) would be spread throughout the universe. Attempts to find this radiation failed for many years, until an unexpected problem in 1965 accidentally detected it.

Gamow is known to many as the author of a number of popular books on science, but he was also a theoretician and experimentalist who liked to throw out ideas right and left, seemingly not caring whether they bore fruit.

Turning his attention to cosmology and the big bang, Gamow noted in 1948 that this dense ball of matter (probably neutrons, he hypothesized) would emit black body radiation, which had been worked out in 1900 by Max Planck. A black body emits radiation at a definable wavelength based on the temperature.

Gamow's two students, Ralph Alpher and Robert Herman, published a paper in 1948 with the calculation for the temperature, and therefore the radiation, of this original ball of matter. The men calculated the temperature to be about 5 degrees above absolute zero, although it took nearly a year for Gamow to agree with this calculation. This radiation is in the microwave range of the electromagnetic spectrum, so it's called the *cosmic microwave background radiation (CMBR)*.

Although this was a successful theoretical breakthrough, it went largely unnoticed at the time. Nobody conducted a serious experiment to look for this radiation, even while Gamow, Alpher, and Herman tried to gain support.

In 1965, a Princeton University team led by Robert Dicke had independently developed the theory and was attempting to test it. Dicke's team failed to discover the CMBR, however, because while they were putting the finishing touches on their equipment, someone else beat them to it.

A few miles away, at New Jersey's Bell Laboratory Holmdell Horn Radio Telescope, Arno Penzias and Robert Wilson were having trouble of their own. Their telescope — which was more sophisticated than Princeton University's — was picking up this horrible static when they attempted to detect radio signals in space. No matter where they pointed the silly thing, they kept getting the same static. The two men even cleaned bird droppings off the telescope, but to no avail. In fact, the static got worse on the unobstructed telescope.

Fortunately, Penzias and Dicke had a mutual friend in astronomer Bernard Burke, and upon discovering the problems the two men had, he introduced them. Penzias and Wilson earned the 1978 Nobel Prize in Physics for accidentally discovering the CMBR (at a temperature of 2.7 degrees above absolute zero — Gamow's calculations had been slightly high).

Forty more years of research has only confirmed the big bang theory, most recently in the picture of the CMBR obtained by the Wilkinson Microwave Anisotropy Probe (WMAP) satellite. The picture obtained by this satellite, shown in Figure 9-3, is like a baby picture of the universe when it was just 380,000 years old (13.7 billion years ago). Before this, the universe was dense enough to be opaque, so no light can be used to look further back than that.

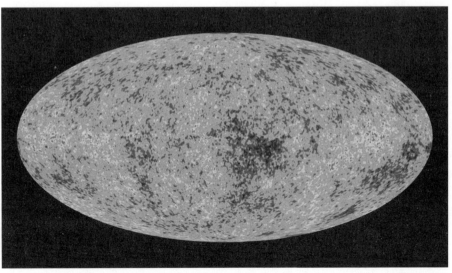

Figure 9-3:
NASA's WMAP satellite image shows a (mostly) uniform cosmic microwave background radiation.

Courtesy of NASA

For more information on the WMAP satellite, check out the official WMAP Web site at NASA's Goddard Space Flight Center, `map.gsfc.nasa.gov`.

Understanding where the chemical elements came from

Both George Gamow and Fred Hoyle, while differing strongly on the big bang theory, were the key figures in determining the process of *stellar nucleosynthesis,* in which atoms are made inside of stars. Gamow theorized that elements were created by the heat of the big bang. Hoyle showed that the heavier elements were actually created by the intense heat of stars and supernovas.

Gamow's original theory was that as the intense heat of the expanding universe cooled, the lightest element, hydrogen, was formed. The energy at this time was still enough to cause hydrogen molecules to interact, perhaps fusing into helium atoms. Estimates show that nearly 75 percent of the visible universe is made up of hydrogen and 25 percent is helium, with the rest of the elements on the periodic table making up only trace amounts on the scale of the entire universe.

This proved to be good, because Gamow couldn't figure out how to cook up many of those heavier elements in the big bang. Hoyle tackled the problem, assuming that if he could make all the elements in stars, then the big bang theory would fail. Hoyle's work on stellar nucleosynthesis was published in 1957.

In Hoyle's nucleosynthesis method, helium and hydrogen gather inside of stars and undergo nuclear fusion. Even this, however, isn't hot enough to make atoms more massive than iron. These heavier elements — zinc, copper, uranium, and many others — are created when massive stars go through their deaths and explode in giant *supernovas.* These supernovas produce enough energy to fuse the protons together into the heavy atomic nucleus.

The elements are then blown out into space by the supernova blast, drifting as clouds of stellar dust. Some of this stellar dust eventually falls together under the influence of gravity to form planets, such as our Earth.

Using Inflation to Solve the Universe's Problems of Flatness and Horizon

In trying to understand the universe, two major problems remained: the *flatness problem* and the *horizon problem.* To solve these two problems, the big bang theory is modified by the *inflation theory,* which states that the universe expanded rapidly shortly after it was created. Today, the principles at the heart of inflation theory have a profound impact on the way that string theory is viewed by many physicists, as becomes clear in Chapter 14.

These two problems can be stated simply as:

- **Horizon problem:** The CMBR is essentially the same temperature in all directions.
- **Flatness problem:** The universe appears to have a flat geometry.

The universe's issues: Too far and too flat

The horizon problem (also sometimes called the *homogeneity problem*) is that no matter which direction you look in the universe, you see basically the same thing (see Figure 9-3). The CMBR temperatures throughout the universe are, to a *very* high level of measurement, almost exactly the same temperature in every direction. This really shouldn't be the case, if you think about it more carefully.

If you look in one direction in space, you're actually looking back in time. The light that hits your eye (or telescope) travels at the speed of light, so it was emitted years ago. This means there's a boundary of 14 billion (or so) light-years in all directions. (The boundary is actually farther because space itself is expanding, but you can ignore that for the purposes of this example.) If there is anything farther away than that, there is no way for it to have ever communicated with us. So you look out with your powerful telescope and can see the CMBR from 14 billion light-years away (call this Point A).

If you now look 14 billion light-years in the opposite direction (call this Point B), you see exactly the same sort of CMBR in that direction. Normally, you'd take this to mean that all the CMBR in the universe has somehow diffused throughout the universe, like heating up an oven. Somehow, the thermal information is communicated between Points A and B.

But Points A and B are 28 billion light-years apart, which means, because no signal can go faster than the speed of light, *there's no way they could have communicated with each other in the entire age of the universe*. How did they become the same temperature if there's no way for heat to transfer between them? This is the horizon problem.

The flatness problem has to do with the geometry of our universe, which appears (especially with recent WMAP evidence) to be a flat geometry, as pictured in Figure 9-2. The matter density and expansion rate of the universe appear to be nearly perfectly balanced, even 14 billion years later when minor variations should have grown drastically. Because this hasn't happened, physicists need an explanation for why the minor variations haven't increased dramatically. Did the variations not exist? Did they not grow into large-scale variations? Did something happen to smooth them out? The flatness problem seeks a reason why the universe has such a seemingly perfectly flat geometry.

Rapid expansion early on holds the solutions

In 1980, astrophysicist Alan Guth proposed the inflation theory to solve the horizon and flatness problems (although later refinements by Andrei Linde, Andreas Albrecht, Paul Steinhardt, and others were required to get it to work). In this model, the early universal expansion accelerated at a rate much faster than we see today.

It turns out that the inflationary theory solves both the flatness problem and horizon problem (at least to the satisfaction of most cosmologists and astrophysicists). The horizon problem is solved because the different regions we see used to be close enough to communicate, but during inflation, space expanded so rapidly that these close regions were spread out to cover all of the visible universe.

The flatness problem is resolved because the act of inflation actually flattens the universe. Picture an uninflated balloon, which can have all kinds of wrinkles and other abnormalities. As the balloon expands, though, the surface smoothes out. According to inflation theory, this happens to the fabric of the universe as well.

In addition to solving the horizon and flatness problems, inflation also provides the seeds for the structure that we see in our universe today. Tiny energy variations during inflation, due simply to quantum uncertainty, become the sources for matter to clump together, eventually becoming galaxies and clusters of galaxies.

One issue with the inflationary theory is that the exact mechanism that would cause — and then turn off — the inflationary period isn't known. Many technical aspects of inflationary theory remain unanswered, though the models include a scalar field called an *inflaton field* and a corresponding theoretical particle called an *inflaton*. Most cosmologists today believe that some form of inflation likely took place in the early universe.

Some variations and alternatives to this model are posed by string theorists and other physicists. Two creators of inflation theory, Andreas Albrecht and Paul J. Steinhart, have worked on alternative theories as well; see Chapter 14 for Steinhart's ekpyrotic theory and Chapter 19 for Albrecht's variable speed of light cosmology.

Dark Matter: The Source of Extra Gravity

Astronomers have discovered that the gravitational effects observed in our universe don't match the amount of matter seen. To account for these differences, it appears that the universe contains a mysterious form of matter that we can't observe, called *dark matter*. Throughout the universe, there's approximately six times as much dark matter as normal visible matter — and string theory may explain where it comes from!

In the 1930s, Swiss astronomer Fritz Zwicky first observed that some galaxies were spinning so fast that the stars in them should fly away from each other. Unfortunately, Zwicky had personality clashes with many in the astronomy community, so his views weren't taken very seriously.

In 1962, astronomer Vera Rubin made the same discoveries and had nearly the same outcome. Though Rubin didn't have the same issues of temperament that Zwicky did, many disregarded her work because she was a woman.

Rubin maintained her focus on the problem and, by 1978, had studied 11 spiral galaxies, all of which (including our own Milky Way) were spinning so fast that the laws of physics said they should fly apart. Together with work from others, this was enough to convince the astronomy community that something strange was happening.

Whatever is holding these galaxies together, observations now indicate that there has to be far more of it than there is the visible matter that makes up the *baryonic matter* that we're used to — the matter that comprises you, this book, this planet, and the stars.

Physicists have made several suggestions about what could make up this dark matter, but so far no one knows for sure. String theorists have some ideas, which you can read about in Chapter 14.

Dark Energy: Pushing the Universe Apart

Einstein's cosmological constant allowed for a uniform repulsive energy throughout the universe. Since Hubble discovered the expansion of the universe, most scientists have believed that the cosmological constant was zero (or possibly slightly negative). Recent findings have indicated that the expansion rate of the universe is actually increasing, meaning that the cosmological constant has a positive value. This repulsive gravity — or *dark energy* — is actually pushing the universe apart. This is one major feature of the universe that string theory may be able to explain.

In 1998, two teams of astronomers announced the same results: Studies of distant *supernovas* (exploding stars) showed that stars looked dimmer than expected. The only way to account for this was if the stars were somehow farther away than expected, but the physicists had already accounted for the expansion of the universe. The explanation eventually found was startling: The rate of expansion of the universe was accelerating.

To explain this, physicists realized that there had to be some sort of repulsive gravity that worked on large scales (see Figure 9-4). On small scales, normal gravity rules, but on larger scales the repulsive gravity force of dark energy seemed to take over. (This doesn't contradict the idea that the universe is flat — but it makes the fact that it is flat, while still expanding, a very unusual and unexpected set of circumstances, which required very narrow parameters on the early conditions of our universe.)

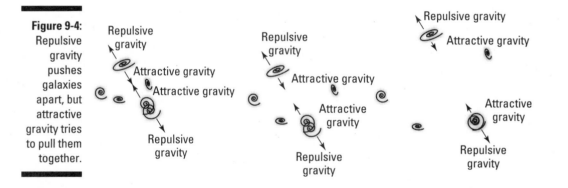

Figure 9-4:
Repulsive gravity pushes galaxies apart, but attractive gravity tries to pull them together.

Repulsive gravity is theorized by inflation theory, but that's a rapid hyper-expansion in the early phases of the universe. Today's expansion due to dark energy may be remnants of the repulsive gravity from inflation, or it may be an entirely distinct phenomenon.

The finding of dark energy (or a positive cosmological constant, which it is roughly similar to) creates major theoretical hurdles, especially considering how weak dark energy is. For years, quantum field theory predicted a huge cosmological constant, but most physicists assumed that some property (such as supersymmetry, which does reduce the cosmological constant value) canceled it out to zero. Instead, the value is non-zero, but differs from theoretical predictions by nearly 120 decimal places! (You can find a more detailed explanation of this discrepancy in Chapter 14.)

In fact, results from the WMAP show that the vast majority of material in our present universe — about 73 percent — is made up of dark energy (remember from relativity that matter and energy are different forms of the same thing: $E = mc^2$, after all). The five-year WMAP data, released in 2008 and shown in Figure 9-5, also allows you to compare the composition of the present universe with the material present in the universe 13.7 billion years ago. The dark energy was a vanishingly small slice of the pie 13.7 billion years ago, but today it eclipses matter and drives the universe's expansion.

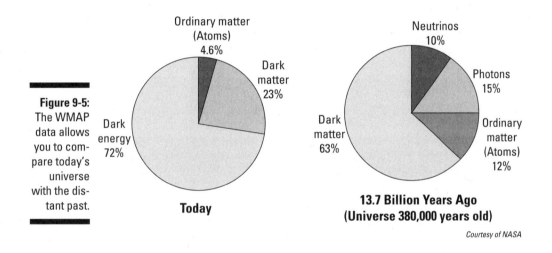

Figure 9-5: The WMAP data allows you to compare today's universe with the distant past.

Courtesy of NASA

The history of the universe is a fascinating topic for study, and trying to understand the meaning of this dark energy is one of the key aspects of modern cosmology. It's also one of the key challenges to modern variations of string theory, as you see in Chapter 11.

Today, many string theorists devote attention to these cosmological mysteries of the universe's origins and evolution because they provide a universal playground on which the ideas of string theory can be explored, potentially at energy levels where string behavior may manifest itself. In Chapters 12, 14, and 15, you discover what behaviors these string theorists might be looking for and what the implications are for the universe.

Stretching the Fabric of Space-Time into a Black Hole

One of the consequences of Einstein's general theory of relativity was a solution in which space-time curved so much that even a beam of light became trapped. These solutions became called *black holes,* and the study of them is one of the most intriguing fields of cosmology. Application of string theory to study black holes is one of the most significant pieces of evidence in favor of string theory.

Black holes are believed to form when stars die and their massive bulk collapses inward, creating intense gravitational fields. No one has "seen" a black hole, but scientists have observed gravitational evidence consistent with predictions about them, so most scientists believe they exist.

What goes on inside a black hole?

According to the general theory of relativity, it's possible that the very fabric of space-time bends an infinite amount. A point with this infinite curvature is called a space-time *singularity.* If you follow space-time back to the big bang, you'd reach a singularity. Singularities also exist inside of black holes, as shown in Figure 9-6.

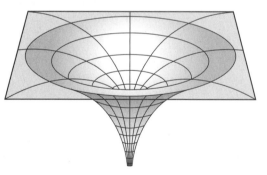

Figure 9-6:
Inside a black hole, space-time stretches to an infinite singularity.

Because general relativity says that the curvature of space-time is equivalent to the force of gravity, the singularity of a black hole has infinite gravity. Any matter going into a black hole would be ripped apart by this intense gravitational energy as it neared the singularity.

For this reason, black holes provide an excellent theoretical testing ground for string theory. Gravity is normally so weak that quantum effects aren't observed, but inside of a black hole, gravity becomes the dominant force at work. A theory of quantum gravity, such as string theory, would explain exactly what happens inside a black hole.

What goes on at the edge of a black hole?

The edge of a black hole is called the *event horizon,* and it represents a barrier that even light can't come out of. If you were to go near the edge of a black hole, relativistic effects take place, including *time dilation.* To an outside observer, it would look like time was slowing down for you, eventually coming to a stop. (You, on the other hand, would notice nothing — until the black hole's intense gravitational forces squished you, of course.)

It was previously believed that things only get sucked into a black hole, but physicist Stephen Hawking famously showed that black holes emit an energy called *Hawking radiation.* (This was proposed in 1974, a year after the equally groundbreaking realization by Israeli Jacob Bekenstein that black holes possessed *entropy* — a thermodynamic measure of disorder in a system. The entropy measures the number of different ways to arrange things in a system.)

Quantum physics predicts that virtual particles are continually created and destroyed, due to quantum fluctuations of energy in the vacuum. Hawking applied this concept to black holes and realized that if such a pair is created near the event horizon, it was possible for one of the particles to get pulled into the black hole while the other one didn't. This would look identical to the black hole emitting radiation. To preserve energy, the particle that fell into the black hole must have negative energy and reduce the overall energy (or mass) of the black hole.

The behavior of black holes is curious in a number of ways, many of them demonstrated by Hawking in the 1970s:

- A black hole's entropy is proportional to the surface area of the black hole (the area of the event horizon), unlike conventional systems where entropy is proportional to volume. This was Bekenstein's discovery.

- If you put more matter into a black hole, it cools down.

- As a black hole emits Hawking radiation, the energy comes from the black hole, so it loses mass. This means the black hole heats up, losing energy (and therefore mass) more quickly.

In other words, Stephen Hawking showed in the mid-1970s that a black hole will evaporate (unless it is "fed" more mass than it loses in energy). He did this by applying principles of quantum physics to a problem of gravity. After the black hole evaporates down to the size of the Planck length, a quantum theory of gravity is needed to explain what happens to it.

Hawking's solution is that the black hole evaporates at that point, emitting a final burst of random energy. This solution results in the so-called *black hole information paradox,* because quantum mechanics doesn't allow information to be lost, but the energy from the evaporation doesn't seem to carry the information about the matter that originally went into the black hole. I discuss this black hole information paradox and its potential resolutions in greater detail in Chapter 14.

Part III
Building String Theory: A Theory of Everything

"What exactly are we saying here?"

In this part . . .

String theory has existed for nearly four decades. It's one of the most unusual scientific theories of all time because it has developed backwards. It began as a theory of particle interactions and failed at that (only to later incorporate the theory that replaced it). It then became a theory of quantum gravity, but made predictions that didn't seem to match reality.

Today string theory has become so complex and has yielded so many unexpected results that its proponents have begun citing this flexibility within the theory as one of its greatest strengths.

This part explains how string theory got its start and how it has transformed over the years. I explain the basic interpretations of key concepts, as well as the ways in which string theorists have been able to adapt to new findings. Finally, I look at some ways that scientists might be able to prove — or disprove — string theory.

Chapter 10

Early Strings and Superstrings: Unearthing the Theory's Beginnings

A year before astronauts set foot on the moon, no one had ever heard of string theory. The concepts at the core of the theory were being neither discussed nor debated. Physicists struggled to complete the Standard Model of particle physics, but had abandoned the hopes of a theory of everything (if they ever had any such hope in the first place).

In other words, no one was looking for strings when physicists found them.

In this chapter, I tell you about the early beginnings of string theory, which quickly failed to do anything the creators expected (or wanted) it to do. Then I explain how, from these humble beginnings, several elements of string theory began to spring up, which drew more and more scientists to pursue it.

Bosonic String Theory: The First String Theory

The first string theory has become known as *bosonic string theory,* and it said that all the particles that physicists have observed are actually the vibration of multidimensional "strings." But the theory had consequences that made it unrealistic to use to describe our reality.

A dedicated group of physicists worked on bosonic string theory between 1968 and the early 1970s, when the development of superstring theory (which said the same thing, but fit reality better) supplanted it. (I explain this superior theory in the later section "Supersymmetry Saves the Day: Superstring Theory.")

Even though bosonic string theory was flawed and incomplete, string theorists occasionally do mathematical work with this model to test new methods and theories before moving on to the more modern superstring models.

Explaining the scattering of particles with early dual resonance models

String theory was born in 1968 as an attempt to explain the scattering of particles (specifically hadrons, like protons and neutrons) within a particle accelerator. Originally, it had nothing to do with strings. These early predecessors of string theory were known as *dual resonance models.*

The initial and final state of particle interactions can be recorded in an array of numbers called an *S-matrix*. At the time, finding a mathematical structure for this S-matrix was considered to be a significant step toward creating a coherent model of particle physics.

Gabriele Veneziano, a physicist at the CERN particle accelerator laboratory, realized that an existing mathematical formula seemed to explain the mathematical structure of the S-matrix. (See the sidebar "Applications of pure mathematics to physics" for more on this formula.) (Physicist Michio Kaku has stated that Mahiko Suzuki, also at CERN, made the same discovery at the same time, but was persuaded by a mentor not to publish it.)

Veneziano's explanation has been called the *dual resonance model,* the *Veneziano amplitude,* or just the *Veneziano model.* The dual resonance model was close to the correct result for how hadrons interacted, but not quite correct. At the time Veneziano developed the model, particle accelerators weren't precise enough to detect the differences between model and reality. (Eventually, it would be shown that the alternative theory of quantum chromodynamics was the correct explanation of hadron behavior, as discussed in Chapter 8.)

Applications of pure mathematics to physics

Physicists frequently find the math they need was created long before it was needed. For example, the equation that physicist Gabriele Veneziano used to explain particle scattering was the Euler beta function, which was discovered in the 1700s by Swiss mathematician Leonhard Euler. Also, when Einstein began to extend special relativity into general relativity, he soon realized that traditional Euclidean geometry wouldn't work. His space had to curve, and Euclid's geometry only described flat surfaces.

Fortunately for Einstein, in the mid-1800s the German mathematician Bernhard Riemann had worked on a form of non-Euclidean geometry (named *Riemannian geometry*). The mathematics that Einstein needed for the general theory of relativity had been created a half century

earlier as an intellectual exercise, with no practical purpose in mind. (As fascinating as revolutionizing the foundations of geometry may be, it was hardly practical.)

This happened several times in the history of string theory. Calabi-Yau manifolds, discussed at the end of this chapter, are one example. Another example is when string theorists were attempting to determine the appropriate number of dimensions to make their theories stable and consistent. A key to this problem came from the journals of Indian mathematical genius Srinivasa Ramanujan (referenced in the film *Good Will Hunting*), who died in 1920. The specific mathematics in this case was a function called the Ramanujan function.

After the dual resonance model was formed, hundreds of theoretical papers were published in attempts to modify the parameters a bit. This was the way theories were approached in physics; after all, an initial guess at a theory is rarely precisely correct and typically requires subtle tweaks — to see how the theory reacts, how much it can be bent and modified, and so on — so that ultimately it fits with the experimental results.

The dual resonance model would have nothing to do with that sort of tinkering — it simply didn't allow for any changes that would still enable it to be valid. The mathematical parameters of the theory were too precisely fixed. Attempts to modify the theory in any way quickly led to a collapse of the entire theory. Like a dagger balanced on its tip, any slight disturbance would send it toppling over. Mathematically, it was locked into a certain set of values. In fact, it has been said by some that the theory had absolutely no adjustable parameters — at least not until it was transformed into an entirely different concept: superstring theory!

This isn't the way theories are supposed to behave. If you have a theory and modify it so the particle mass, for example, changes a bit, the theory shouldn't collapse — it should just give you a different result.

When a theory can't be modified, there are only two possible reasons: either it's completely wrong or it's completely right! For several years, dual resonance models looked like they might be completely right, so physicists continued to ponder what they might mean.

Exploring the first physical model: Particles as strings

The basic physical interpretation of string theory was as vibrating strings. As the strings, each representing a particle, collided with each other, the S-matrix described the result.

 Consider this very informal way of looking at string theory, shown in Figure 10-1. Each particle is composed of a vibrating string. In the case of a proton, there are three quark strings. When these three strings come in contact, they bond together to form a proton. So the proton is created by the interaction of the three quark strings touching each other. The proton is kind of a knot within the strings.

Figure 10-1:
Most people think of particles as solid spheres. In string theory, scientists view them as vibrating strings instead.

 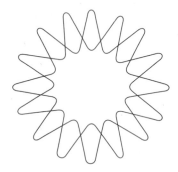

What are these strings like? The strings described were almost like rubber bands. There is a certain "springiness" to them. A phrase that I think describes them well is "filaments of energy" (as string theorist Brian Greene and others have called them). Though most people think of particles as balls of matter, physicists have long thought of them as little bundles of waves (called *wave packets*), which is in line with describing them as strings. (In some other situations, physicists can treat particles as having no size whatsoever, but this is a simplification to make the math and theory more manageable. The way physicists treat particles depends a lot on the situation they're working with.)

This interpretation was put forth independently by Yoichiro Nambu, Holger Nielsen, and Leonard Susskind in 1970, earning all three men positions as founders of string theory.

According to Einstein's work, mass was a form of energy, an insight demonstrated dramatically by the creation of the atomic bomb. Quantum theory showed physicists that matter was represented by the mathematics of wave mechanics, so even a particle had a wavelength associated with it.

In string theory, matter again takes on a new form. Particles of different types are different vibrational modes of these fundamental entities: energetic rubber bands, or *strings*. (Classical vibrations and strings are discussed in Chapter 5.) In essence, the more the string vibrates, the more energy (and therefore mass) it possesses.

Through all the transformations that string theory has undergone in the years since its discovery, this central concept remains (fairly) constant, although in recent years new objects in addition to strings have been introduced (which I explain in Chapter 11 when I discuss branes).

The basic physical model couldn't have been simpler: The particles and forces in nature are really interactions between vibrating strings of energy.

Bosonic string theory loses out to the Standard Model

The dual resonance model was created for the express purpose of explaining the S-matrix particle scattering, which was now explained in terms of the Standard Model of particle physics — gauge fields and quantum chromodynamics. (See Chapter 8 for more on these concepts.) There was no point to string theory in light of the success of the Standard Model.

Also, as the measurements of experiments in particle accelerators got more precise, it became clear that dual resonance models were only approximately correct. In 1969, physicists showed that Veneziano had discovered only the first term in an infinite series of terms. Although this term was the most important, it still wasn't complete. The theory appeared to need some further refinement to match the results perfectly.

Terms could be added (which Michio Kaku did in 1972), correcting for the different ways that the strings could collide, but it made the theory less elegant. There were growing indications that string theory might not work the way everyone had thought it would and that, indeed, quantum chromodynamics explained the behavior of the particle collisions better.

The early string theorists had therefore spent a lot of time giving meaning to a theory that seemed to (almost) accurately predict the S-matrix, only to find that the majority of particle physicists weren't interested in it. It had to be very frustrating to have such an elegant model that was quickly falling into obscurity.

But a few string theorists weren't about to give up on it quite yet.

Why Bosonic String Theory Doesn't Describe Our Universe

By 1974, bosonic string theory was quickly becoming a mathematical mess, and attempts to make the theory mathematically consistent caused more trouble for the model than it had already. Playing with the math introduced four conditions that should have, by all rights, spelled the end of the early string theory:

- Massless particles
- Tachyons, which move faster than the speed of light
- Fermions, such as electrons, can't exist
- 25 spatial dimensions

The cause of these problems was a reasonable constraint built into string theory. No matter what else string theory did, it needed to be consistent with existing physics — namely special relativity and quantum theory.

The Standard Model of particle physics was consistent with both theories (though it still had trouble reconciling with general relativity), so string theory also had to be consistent with both. If it violated a half century of established physics, there was no way it could be a viable theory.

Physicists eventually found ways to modify the theory to be consistent with these existing physical laws. Unfortunately, these modifications resulted in the four problematic features outlined in the bulleted list. It wasn't just that these features were possible, but that they were now seemingly essential components of the theory.

Massless particles

One side effect of creating a consistent string theory is that it had to contain certain objects that can never be brought to rest. Because mass is a measure of an object while it's at rest, these sorts of particles are called *massless particles*. This would be a major problem for string theory if the massless particles predicted didn't really exist.

Overall, though, this wasn't a terribly disturbing problem because scientists know for certain that at least one particle exists only in a state of motion: the photon. (The gluon, though not known for certain at the time, is also a massless particle.)

Under the Standard Model of particle physics at the time, it was believed that a particle called the *neutrino* might have a mass of zero. (Today we know that the neutrino's mass is slightly higher than zero.)

There was also one other possible massless particle: the graviton. The *graviton* is the theoretical gauge boson that could be responsible for the force of gravity under quantum field theory.

The existence of massless particles in string theory was unfortunate, but it was a surmountable problem. String theorists needed to uncover the properties of massless particles and prove that their properties were consistent with the known universe.

Tachyons

A bigger problem than massless particles was the *tachyon,* a particle predicted by bosonic string theory that travels faster than the speed of light. Under a consistent bosonic string theory, the mathematical formulas demanded that tachyons exist, but the presence of tachyons in a theory represents a fundamental instability in the theory. Solutions that contain tachyons will always decay into another, lower energy solution — possibly in a never-ending cycle. For this reason, physicists don't believe that tachyons really exist, even if a theory initially looks like it contains such particles.

Strictly speaking, Einstein's theory of relativity doesn't absolutely forbid an object from traveling faster than the speed of light. What it says is that it would require an infinite amount of energy for an object to *accelerate* to the speed of light. Therefore, in a sense, the tachyon would still be consistent with relativity, because it would *always* be moving faster than the speed of light (and wouldn't ever have to accelerate to that speed).

Mathematically, when calculating a tachyon's mass and energy using relativity, it would contain imaginary numbers. (An *imaginary number* is the square root of a negative number.)

This was exactly how string theory equations predicted the tachyon: They were consistent only if particles with imaginary mass existed. But what is imaginary mass? What is an imaginary energy? These physical impossibilities give rise to the problems with tachyons.

The presence of tachyons is in no way unique to bosonic string theory. For example, the Standard Model contains a certain vacuum in which the Higgs boson is actually a type of tachyon as well. In this case, the theory isn't inconsistent; it just means that the solution that was applied wasn't a stable solution. It's like trying to place a ball at the top of a hill — any slight movement will cause the ball to roll into a nearby valley. Similarly, this tachyon solution decays into a stable solution without the tachyons.

Unfortunately, in the case of bosonic string theory, there was no clear way to figure out what happened during the decay, or even if the solution ended up in a stable solution after decaying into a lower energy state.

With all of these problems, physicists don't view these tachyons as actual particles that exist, but rather as mathematical artifacts that fall out of the theory as a sign of certain types of inherent instabilities. Any solution that contains tachyons quickly decays due to these instabilities.

Some physicists (and science fiction authors) have explored notions of how to treat tachyons as actual particles, a speculative concept that will come up briefly in Chapter 16. But for now, just know that tachyons were one of the things that made physicists decide, at the time, that bosonic string theory was a failure.

No electrons allowed

The real flaw in bosonic string theory was the one that it's named after. The theory predicted only the existence of bosons, not fermions. Photons could exist, but not quarks or electrons.

Every elementary particle observed in nature has a property called a *spin,* which is either an integer value (–1, 0, 1, 2, and so on) or a half-integer value (–½, ½, and so on). Particles with integer spins are *bosons,* and particles with half-integer spins are *fermions.* One key finding of particle physics is that all particles fall into one of these two categories.

For string theory to apply to the real world it had to include both types of particles, and the original formulation didn't. The only particles allowed under the first model of string theory were bosons. This is why it would come to be known to physicists as the *bosonic string theory.*

25 space dimensions, plus 1 of time

Dimensions are the pieces of information needed to determine a precise point in space. (Dimensions are generally thought of in terms of up/down, left/right, forward/backward.) In 1974, Claude Lovelace discovered that bosonic string theory could only be physically consistent if it were formulated in 25 spatial dimensions (Chapter 13 delves into the idea of the additional dimensions in more depth), but so far as anyone knows, we only have three spatial dimensions!

Relativity treats space and time as a continuum of coordinates, so this means that the universe has a total of 26 dimensions in string theory, as opposed to the four dimensions it possesses under Einstein's special and general relativity theories.

It's unusual that this requirement would be implicit in the theory. Einstein's relativity has three spatial dimensions and one time dimension because those are the conditions used to create the theory. He didn't begin working on relativity and just happen to stumble upon three spatial dimensions, but rather intentionally built it into the theory from the beginning. If he'd wanted a 2-dimensional or 5-dimensional relativity, he could have built the theory to work in those dimensions.

With bosonic string theory, the equations actually demanded a certain number of dimensions to be mathematically consistent. The theory falls apart in any other number of dimensions!

The reason for extra dimensions

The reason for these extra dimensions can be seen by analogy. Consider a long, loose spring (like a Slinky), which is flexible and elastic, similar to the strings of string theory. If you lay the spring in a straight line flat on the floor and pull it outward, waves move along the length of the spring. These are called *longitudinal waves* and are similar to the way sound waves move through the air.

The key thing is that these waves, or vibrations, move only back and forth along the length of the spring. In other words, they're 1-dimensional waves.

Now imagine that the spring stays on the floor, but someone holds each end. Each person can move the ends of the spring anywhere they want, so long as it stays on the floor. They can move it left and right, or back and forth, or some combination of the two. As the ends of the spring move in this way, the waves that are generated require two dimensions to describe the motion.

Finally, imagine that each person has an end of the spring but can move it anywhere — left or right, back or forth, and up or down. The waves generated by the spring require three dimensions to explain the motion. Trying to use 2-dimensional or 1-dimensional equations to explain the motion wouldn't make sense.

In an analogous way, bosonic string theory required 25 spatial dimensions so the symmetries of the strings could be fully consistent. (*Conformal symmetry* is the exact name of the type of symmetry in string theory that requires this number of dimensions.) If the physicists left out any of those dimensions, it made about as much sense as trying to analyze the 3-dimensional spring in only one dimension . . . which is to say, none at all.

Dealing with the extra dimensions

The physical conception of these extra dimensions was (and still is) the hardest part of the theory to comprehend. Everyone can understand three spatial dimensions and a time dimension. Give me a latitude, longitude, altitude, and time, and I can meet you anywhere on the planet. You can measure height, width, and length, and you experience the passage of time, so you have a regular familiarity with what those dimensions represent.

What about the other 22 spatial dimensions? It was clear that these dimensions had to be hidden somehow. The Kaluza-Klein theory predicted that extra dimensions were rolled up, but rolling them up in precisely the right way to achieve results that made sense was difficult. This was achieved for string theory in the mid-1980s through the use of Calabi-Yau manifolds, as I discuss later in this chapter.

No one has any direct experience with these strange other dimensions. For the idea to come out of the symmetry relationships associated with a relatively obscure new theoretical physics conjecture certainly didn't offer much motivation for physicists to accept it. And for more than a decade, most physicists didn't.

Supersymmetry Saves the Day: Superstring Theory

Despite bosonic string theory's apparent failures, some brave physicists stayed committed to their work. Why? Well, physicists can be a passionate bunch (nearly obsessive, some might say). Another reason was that by the time these problems were fully realized, many string theorists had already moved on from bosonic string theory anyway.

With the development of *supersymmetry* in 1971, which allows for bosons and fermions to coexist, string theorists were able to develop *supersymmetric string theory,* or, for short, *superstring theory,* which took care of the major problems that destroyed bosonic string theory. This work opened up whole new possibilities for string theory.

Almost every time you hear or read the phrase "string theory," the person probably really means "superstring theory." Since the discovery of supersymmetry, it has been applied to virtually all forms of string theory. The only string theory that really has nothing to do with supersymmetry is bosonic string theory, which was created before supersymmetry. For all practical discussion purposes (with anyone who isn't a theoretical physicist), "string theory" and "superstring theory" are the same term.

Fermions and bosons coexist . . . sort of

Symmetries exist throughout physics. A *symmetry* in physics is basically any situation where two properties can be swapped throughout the system and the results are precisely the same.

The notion of symmetry was picked up by Pierre Ramond in 1970, followed by the work of John Schwarz and Andre Neveu in 1971, to give hope to string theorists. Using two different techniques, they showed that bosonic string theory could be generalized in another way to obtain non-integer spins. Not only were the spins non-integer, but they were precisely half-integer spins, which characterize the fermion. No spin ¼ particles showed up in the theory, which is good because they don't exist in nature.

Including fermions into the model meant introducing a powerful new symmetry between fermions and bosons, called supersymmetry. *Supersymmetry* can be summarized as

- ✔ Every boson is related to a corresponding fermion.
- ✔ Every fermion is related to a corresponding boson.

In Chapter 11, I discuss the reasons to believe that supersymmetry is true, as well as ways that it can be proved. For now, it's enough to know that it's needed to make string theory work.

Who discovered supersymmetry?

The origins of supersymmetry are a bit confusing, because it was discovered around the same time by four separate groups.

In 1971, Russians Evgeny Likhtman and Yuri Golfand created a consistent theory containing supersymmetry. A year later, they were followed by two more Russians, Vladimir Akulov and Dmitri Volkov. These theories were in only two dimensions, however.

Due to the Cold War, communication between Russia and the non-communist world wasn't very good, so many physicists didn't hear about the Russian work. European physicists Julius Wess and Bruno Zumino were able to create a 4-dimensional supersymmetric quantum theory in 1973, probably aware of the Russian work. Theirs was noticed by the Western physics community at large.

Then, of course, we have Pierre Ramond, John Schwarz, and Andre Neveu, who developed supersymmetry in 1970 and 1971, in the context of their superstring theories. It was only on later analysis that physicists realized their work and the later work hypothesized the same relationships.

Many physicists consider this repeated discovery as a good indication that there's probably something to the idea of supersymmetry in nature, even if string theory itself doesn't prove to be correct.

Of course, as you'll anticipate if you're looking for trends in the story of string theory, things didn't quite fall out right. Fermions and bosons have very different properties, so getting them to change places without affecting the possible outcomes of an experiment isn't easy.

Physicists know about a number of bosons and fermions, but when they began looking at the properties of the theory, they found that the correspondence didn't exist between known particles. A photon (which is a boson) doesn't appear to be linked by supersymmetry with any of the known fermions.

Fortunately for theoretical physicists, this messy experimental fact was seen as only a minor obstacle. They turned to a method that has worked for theorists since the dawn of time. If you can't find evidence of your theory, hypothesize it!

Double your particle fun: Supersymmetry hypothesizes superpartners

Under supersymmetry, the corresponding bosons and fermions are called *superpartners*. The superpartner of a standard particle is called a *sparticle*.

Because none of the existing particles are superpartners, this means that if supersymmetry is true, there are twice as many particles as we currently know about. For every standard particle, a sparticle that has never been detected experimentally must exist. The detection of sparticles will be one of the key pieces of evidence the Large Hadron Collider will look for.

If I mention a strangely named particle that you've never run into, it's probably a sparticle. Because supersymmetry introduces so many new particles, it's important to keep them straight. Physicists have introduced a Dr. Seuss-like naming convention to identify the hypothetical new particles:

- ✔ The superpartner of a fermion begins with an "s" before the standard particle name; so the superpartner of an "electron" is the "selectron," and the superpartner of the "quark" is the "squark."

- ✔ The superpartner of a boson ends in an "–ino," so the superpartner of a "photon" is the "photino" and of the "graviton" is the "gravitino."

Table 10-1 shows the names of standard particles and their corresponding superpartner.

Table 10-1	Some Superpartner Names
Standard Particle	*Superpartner*
Lepton	Slepton
Muon	Smuon
Neutrino	Sneutrino
Top Quark	Stop Squark
Gluon	Gluino
Higgs boson	Higgsino
W boson	Wino
Z boson	Zino

Even though there is an elementary superpartner called a "sneutrino," there exists no elementary particle called a "sneutron."

Some problems get fixed, but the dimension problem remains

The introduction of supersymmetry into string theory helped with some of the major problems of bosonic string theory. Fermions now existed within the theory, which had been the biggest problem. Tachyons vanished from superstring theory. Massless particles were still present in the theory, but weren't seen as a major issue. Even the dimensional problem improved, dropping from 26 space-time dimensions down to a mere ten.

The supersymmetry solution was elegant. Bosons — the photon, graviton, Z, and W bosons — are units of force. Fermions — the electron, quarks, and neutrinos — are units of matter. Supersymmetry created a new symmetry, one between matter and forces.

In 1972, Andre Neveu and Joel Scherk resolved the massless particle issue by showing that string vibrational states could correspond to the gauge bosons, such as the massless photon.

The dimensional problem remained, although it was better than it had been. Instead of 25 spatial dimensions, superstring theory became consistent with a "mere" nine spatial dimensions (plus one time dimension, for a total of ten dimensions). Many string theorists of the day believed this was still too many dimensions to work with, so they abandoned the theory for other lines of research.

One physicist who turned his back on string theory was Michio Kaku, one of today's most vocal advocates of string theory. Kaku's PhD thesis involved completing all the terms in the Veneziano model's infinite series. He'd created a field theory of strings, so he was working in the thick of string theory. Still, he abandoned work on superstring theory, believing that there was no way it could be a valid theory. That's how serious the dimensional problem was.

For the handful of people who remained dedicated to string theory after 1974, they faced serious issues about how to proceed. With the exception of the dimensional problem, they had resolved nearly all the issues with bosonic string theory by transforming it into superstring theory.

The only question was what to do with it.

Supersymmetry and Quantum Gravity in the Disco Era

By 1974, the Standard Model had become the theoretical explanation of particle physics and was being confirmed in experiment after experiment. With a stable foundation, theoretical physicists now looked for new worlds to conquer, and many decided to tackle the same problem that had vexed Albert Einstein for the last decades of his life: quantum gravity.

Also as a consequence of the Standard Model's success, string theory wasn't needed to explain particle physics. Instead, almost by accident, string theorists began to realize that string theory might just be the very theory that would solve the problem of quantum gravity.

The graviton is found hiding in string theory

The graviton is a particle that, under predictions from unified field theory, would mediate the gravitational force (see Chapter 2 for more on the graviton). In a very real sense, the graviton *is* the force of gravity. One major finding of string theory was that it not only includes the graviton, but requires its existence as one of the massless particles discussed earlier in this chapter.

In 1974, Joel Scherk and John Schwarz demonstrated that a spin-2 massless particle in superstring theory could actually be the graviton. This particle was represented by a closed string (which formed a loop), as opposed to an open string, where the ends are loose. Both sorts of strings are demonstrated in Figure 10-2.

Figure 10-2:
String theory allows for open and closed strings. Open strings are optional, but closed strings have to exist.

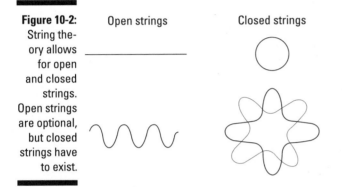

Open strings   Closed strings

String theory demands that these closed strings must exist, though open strings may or may not exist. Some versions of string theory are perfectly mathematically consistent but contain *only* the closed strings. No theory contains only open strings, because if you have open strings, you can construct a situation where the ends of the strings meet each other and, voilà, a closed string exists. (Cutting closed strings to get open strings isn't always allowed.)

From a theoretical standpoint, this was astounding (in a good way). Instead of trying to shoehorn gravity into the theory, the graviton fell out as a natural consequence. If superstring theory was the fundamental law of nature, then it required the existence of gravity in a way that no other proposed theory had ever done!

Immediately, it became clear to Schwarz and Scherk that they had a potential candidate for quantum gravity on their hands.

Even while everyone else was fleeing from the multiple dimensions their theory predicted, Scherk and Schwarz became more convinced than ever that they were on the right track.

The other supersymmetric gravity theory: Supergravity

Supergravity is the name for theories that attempt to apply supersymmetry directly to the theory of gravity without the use of string theory. Throughout the late 1970s, this work proceeded at a faster pace than string theory, mainly because it was popular while the string theory camp had become a ghost town. Supergravity theories prove important in the later development of M-theory, which I cover in Chapter 11.

In 1976, Daniel Freedman, Sergio Ferrara, and Peter van Nieuwenhuizen applied supersymmetry to Einstein's theory of gravity, resulting in a theory of supergravity. They did this by introducing the superpartner of the graviton, the gravitino, into the theory of general relativity.

Building on this work, Eugene Cremmer, Joel Scherk, and Bernard Julia were able to show in 1978 that supergravity could be written, in its most general form, as an 11-dimensional theory. Supergravity theories with more than 11 dimensions fell apart.

Supergravity ultimately fell prey to the mathematical inconsistencies that plagued most quantum gravity theories (it worked fine as a classical theory, so long as you kept it away from the quantum realm), leaving room for superstring theory to rise again in the mid-1980s, but it didn't go away completely. I return to the idea of the 11-dimensional supergravity theory in Chapter 11.

String theorists don't get no respect

During the late 1970s, string theorists were finding it hard to be taken seriously, let alone find secure academic work. String theorists' search for respect in the field of physics reminds me of a young Einstein working in the Bern patent office, denied job after job while he thought about mass and energy.

There had been earlier issues in getting recognition for string theory work. The journal *Physics Review Letters* didn't consider Susskind's 1970 work — interpreting the dual resonance model as vibrating strings — significant enough to publish. Susskind himself tells how physics giant Murray Gell-Mann laughed at him for mentioning string theory in 1970. (The story ends well, with Gell-Mann expressing interest in the theory in 1972.)

As the decade progressed, two of the major forces behind string theory would run into hurdle after hurdle in getting a secure professorship. John Schwarz had been denied tenure at Princeton in 1972 and spent the next 12 years at CalTech in a temporary position, never sure if the funding for his job would be renewed. Pierre Ramond, who had discovered supersymmetry and helped rescue string theory from oblivion, was denied tenure at Yale in 1976.

Against the backdrop of professional uncertainty, the few string theorists continued their work through the late 1970s and early 1980s, helping deal with some of the extra dimensional hurdles in supergravity and other theories, until the day came when the tables turned and they were able to lay claim to the high ground of theoretical physics.

A Theory of Everything: The First Superstring Revolution

The year 1984 is marked by many as the start of "the first superstring revolution." The major finding that sparked the revolution was the proof that string theory contained no anomalies, unlike many of the quantum gravity theories, including supergravity, studied during the 1970s.

For nearly a decade, John Schwarz had been working on showing that superstring theory could be a quantum theory of gravity. His major partner in this, Joel Scherk, had died in 1980, a tragic blow to the cause. By 1983, Schwarz was working with Michael Green, one of the few individuals who had been persuaded to work on string theory during that time.

Typically, two major problems arose in theories of quantum gravity: anomalies and infinities. Neither is a good sign for a scientific theory.

- ✔ **Infinities** occur when values, such as energy, probability, or curvature, begin increasing rapidly to an infinite value.

- ✔ **Anomalies** are cases where quantum mechanical processes can violate a symmetry that is supposed to be preserved.

Superstring theory was actually pretty good at avoiding infinities.

One simplification that allows you to understand, in very general terms, how superstring theory avoids infinities is that the distance value never quite reaches zero. Dividing by zero (or a value that can get arbitrarily close to zero) is the mathematical operation that results in an infinity. Because the strings have a tiny bit of length (I call it L), the distance never gets smaller than L, and so the gravitational force is obtained by dividing by a number that never gets smaller than L^2. This means that the gravitational force will never explode up to infinity, as happens when the distance approaches zero without a limit.

String theory also had no anomalies (at least under certain specific conditions), as Schwarz and Green proved in 1984. They showed that certain 10-dimensional versions of superstring theory had exactly the constraints needed to cancel out all anomalies.

This changed the whole landscape of theoretical physics. For a decade, superstring theory had been ignored while every other method of creating a quantum theory of gravity collapsed in upon itself under infinities and anomalies. Now this discarded theory had risen from the ashes like a mathematical phoenix — both finite and anomaly free.

Theorists began to think that superstring theory had the potential to unify all the forces of nature under one simple set of physical laws with an elegant model in which everything consisted of different energy levels of vibrating strings. It was the ideal that had eluded Einstein: a fundamental theory of all natural law that explained all observed phenomena.

But We've Got Five Theories!

In the wake of 1984's superstring revolution, work on string theory reached a fever pitch. If anything, it proved a little too successful. It turned out that instead of one superstring theory to explain the universe, there were five, given the colorful names

- Type I
- Type IIA
- Type IIB
- Type HO
- Type HE

And, once again, each one *almost* matched our world . . . but not quite.

By the time the decade ended, physicists had developed and dismissed many variants of string theory in the hopes of finding the one true formulation of the theory.

Instead of one formulation, though, five distinct versions of string theory proved to be self-consistent. Each had some properties that made physicists think it would reflect the physical reality of our world — and some properties that are clearly not true in our universe.

The distinctions between these theories are mathematically sophisticated. I introduce their names and basic definitions mainly because of the key role they play in M-theory, which I introduce in Chapter 11.

Type 1 string theory

Type I string theory involves both open and closed strings. It contains a form of symmetry that's mathematically designated as a symmetry group called O(32). (I'll try to make that the most mathematics you need to know related to symmetry groups.)

Type IIA string theory

Type IIA string theory involves closed strings where the vibrational patterns are symmetrical, regardless of whether they travel left or right along the closed string. Type IIA open strings are attached to structures called D-branes (which I discuss in greater detail in Chapter 11) with an odd number of dimensions.

Type IIB string theory

Type IIB string theory involves closed strings where the vibrational patterns are asymmetrical, depending upon whether they travel left or right along the closed string. Type IIB open strings are attached to D-branes (discovered in 1995 and covered in Chapter 11) with an even number of dimensions.

Two strings in one: Heterotic strings

A new form of string theory, called *heterotic string theory,* was discovered in 1985 by the Princeton team of David Gross, Jeff Harvey, Emil Martinec, and Ryan Rohm. This version of string theory sometimes acted like bosonic string theory and sometimes acted like superstring theory.

A distinction of the heterotic string is that the string vibrations in different directions resulted in different behaviors. "Left-moving" vibrations resembled the old bosonic string, while "right-moving" vibrations resembled the Type II strings. The heterotic string seemed to contain exactly the properties that Green and Schwarz needed to cancel out anomalies within the theory.

It was ultimately shown that only two mathematical symmetry groups could be applied to heterotic string theory, which resulted in stable theories in ten dimensions — O(32) symmetry and $E_8 \times E_8$ symmetry. These two groups gave rise to the names Type HO and Type HE string theory.

Type HO string theory

Type HO is a form of heterotic string theory. The name comes from the longer name Heterotic O(32) string theory, which describes the symmetry group of the theory. It contains only closed strings whose right-moving vibrations resemble the Type II strings and whose left-moving vibrations resemble the bosonic strings. The similar theory, Type HE, has subtle but important mathematical differences regarding the symmetry group.

Type HE string theory

Type HE is another form of heterotic string theory, based on a different symmetry group from the Type HO theory. The name comes from the longer name Heterotic $E_8 \times E_8$ string theory, based on the symmetry group of the theory. It also contains only closed strings whose right-moving vibrations resemble the Type II strings and whose left-moving vibrations resemble the bosonic strings.

How to Fold Space: Introducing Calabi-Yau Manifolds

The problem of extra dimensions continued to plague string theory, but these were solved by introducing the idea of *compactification,* in which the extra dimensions curl up around each other, growing so tiny that they're extremely hard to detect. The mathematics about how this might be achieved had already been developed in the form of complex *Calabi-Yau manifolds,* an example of which is shown in Figure 10-3. The problem is that string theory offers no real way to determine exactly which of the many Calabi-Yau manifolds is right!

When the extra dimensions were first discovered in the 1970s, it was clear that they must be hidden in some way. After all, we certainly don't see more than three spatial dimensions.

One suggestion was the one that had been proposed by Kaluza and Klein a half century earlier: The dimensions could be curled up into a very small size.

Early attempts to curl up these extra dimensions ran into problems because they tended to retain the symmetry between left- and right-handed particles (called *parity* by physicists), which isn't always retained in nature. This violation is crucial in understanding the operation of the weak nuclear force.

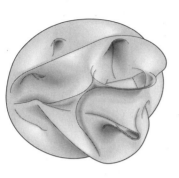

For string theory to work, there had to be a way to compactify the extra six dimensions while still retaining a distinction between the left-handed and right-handed particles.

In 1985, the Calabi-Yau manifolds (created for other purposes years earlier by mathematicians Eugenio Calabi and Shing-Tung Yau) were used by Edward Witten, Philip Candelas, Gary Horowitz, and Andrew Strominger to compactify the extra six space dimensions in just the right way. These manifolds not only preserved the handedness of the particles, but they also preserved super-symmetry just enough to replicate certain aspects of the Standard Model.

One benefit of the Calabi-Yau manifolds was that the geometry of the folded dimensions gives rise to different types of observable particles in our universe. If the Calabi-Yau shape has three holes (or rather higher-dimensional analogs of holes), three families of particles will be predicted by the Standard Model of particle physics. (Obviously, by extension, a shape with five holes will have five families, but physicists are only concerned with the three families of particles that they know exist in this universe.)

Unfortunately, there are tens of thousands of possible Calabi-Yau manifolds for six dimensions, and string theory offers no reasonable means of determining which is the right one. For that matter, even if physicists could determine which one was the right one, they'd still want to answer the question of why the universe folded up the extra six dimensions in that particular configuration.

When Calabi-Yau manifolds were first discovered, it was hoped by some vocal members of the string theory community that one specific manifold would fall out as the right one. This hasn't proved to be the case, and this is what many string theorists would have expected in the first place — that the specific Calabi-Yau manifold is a quantity that has to be determined by experiment. In fact, it's now known that some other geometries for folded spaces can also maintain the needed properties. I talk about the implications of this folded space — what it could really mean — in Chapters 13 and 14.

String Theory Loses Steam

The rising tide of string theory research couldn't last forever, and by the early 1990s some were giving up any hope of finding one single theory. Just as the earlier introduction of multiple dimensions had warded off new physicists, the rise of so many distinct yet consistent versions of string theory gave many physicists pause. Physicists who were motivated purely by the drive to find a quick and easy "theory of everything" began turning away from string theory when it became clear that there was nothing quick and easy about it. As the easier problems got solved and only the harder ones remained, the truly dedicated retained the motivation to work through the complications.

In 1995, a second string theory revolution would come along, with the rise of new insights that would help convince even many of the skeptics that work on string theory would ultimately bear significant fruit. That second revolution is the topic of Chapter 11.

Chapter 11

M-Theory and Beyond: Bringing String Theory Together

· ·

In This Chapter

▶ M-theory re-energizes the movement

▶ Thinking about branes

▶ Overcoming the conundrum of dark energy

▶ So many string theories, why pick just one?

· ·

The last chapter ended with five versions of string theories. Theorists continued their work, but were uncertain how to take these findings. A new insight was needed to generate further progress in the field.

In this chapter, I explain how that insight came about in the form of M-theory, which unified these string theories into one theory. I discuss how string theory was expanded to include objects with more than one dimension, called branes. I introduce some possible insights that may help explain what M-theory is trying to describe. I show how the discovery of dark energy, unpredicted by string theory, has complicated string theory, as well as introduced a large number of possible correct solutions to the theories. Finally, I examine how some physicists have used the anthropic principle to try to give meaning to this landscape of string theories.

Introducing the Unifying Theory: M-Theory

At a conference in 1995, physicist Edward Witten proposed a bold resolution to the problem of five distinct string theories. In his theory, based on newly discovered dualities, each of the existing theories was a special case of one overarching string theory, which he enigmatically called *M-theory*. One of the key concepts required for M-theory was the introduction of branes (short for membranes) into string theory. *Branes* are fundamental objects in string theory that have more than one dimension.

Witten didn't thoroughly explain the true meaning of the name M-theory, leaving it as something that each person can define for himself. There are several possibilities for what the "M" could stand for: membrane, magic, mother, mystery, or matrix. Witten probably took the "M" from membrane because those featured so prominently in the theory, but he didn't want to commit himself to requiring them so early in the development of the new theory.

Although Witten didn't propose a complete version of M-theory (in fact, we're still waiting on one), he did outline certain defining traits that M-theory would have:

- 11 dimensions (10 space dimensions plus 1 time dimension)
- Dualities that result in the five existing string theories all being different explanations of the same physical reality
- Branes — like strings, but with more than one dimension

Translating one string theory into another: Duality

The core of M-theory is the idea that each of the five string theories introduced in Chapter 10 is actually a variation on one theory. This new theory — M-theory — is an 11-dimensional theory that allows for each of the existing theories (which are 10-dimensional) to be equivalent if you make certain assumptions about the geometry of the space involved.

The basis for this suggestion was the understanding of dualities that were being recognized among the various string theories. A *duality* occurs when you can look at the same phenomenon in two distinct ways, taking one theory and mapping it to another theory. In a sense, the two theories are equivalent. By the mid-1990s, growing evidence showed that at least two dualities existed between the various string theories; they were called *T-duality* and *S-duality*.

These dualities were based on earlier dualities conjectured in 1977 by Claus Montonen and David Olive. In the early 1990s, Indian physicist Ashoke Sen and Israeli-born physicist Nathan Seiberg did work that expanded on the notions of these dualities. Witten drew upon this work, as well as more recent work by Chris Hull, Paul Townsend, and Witten himself, to present M-theory.

Topology: The mathematics of folding space

The study of topology allows you to study mathematical spaces by eliminating all details from the space except for certain sets of properties that you care about. Two spaces are topologically equivalent if they share these properties, even if they differ in other details. Certain actions may be more easily performed on one of the spaces than the other. You then perform actions on that space and can work backward to find the resulting effect on the topologically equivalent space. It can be far easier than trying to perform these actions on the original space directly.

One of the key components of topology is the study of how different topological spaces relate to each other. Much of the time, these different spaces involve some sort of manipulation of the space, which is what adds the complexity. If this manipulation can be performed without breaking or reconnecting the space in a new way, the two spaces are topologically equivalent.

To picture this, imagine a donut (or torus) of clay that you slowly and meticulously recraft into the shape of a coffee mug. The hole in the center of the donut never has to be broken in order to be turned into the handle of the coffee mug. On the other hand, if you start with a donut, there's no way to turn it into a pretzel without introducing breaks into the space — a donut and a pretzel are topologically distinct.

Topological duality: T-duality

One of the dualities discovered at the time was called *T-duality*, which refers to either *topological duality* or *toroidal duality*, depending on whom you ask. (*Toroidal* is a reference to the simplest case, which is a *torus*, or donut shape. *Topological* is a precise way of defining the structure of that space, as explained in the nearby sidebar "Topology: The mathematics of folding space." In some cases the T-duality has nothing to do with a torus, and in other cases, it's not topological.) The T-duality related the Type II string theories to each other and the heterotic string theories to each other, indicating that they were different manifestations of the same fundamental theory.

In the T-duality, you have a dimension that is compactified into a circle (of radius R), so the space becomes something like a cylinder. It's possible for a closed string to wind around the cylinder, like thread on a spindle. (This means that both the dimension and the string have radius R.) The number of times the closed string winds around the cylinder is called the *winding number*. You have a second number that represents the momentum of the closed string.

Here's where things get interesting. For certain types of string theory, if you wrap one string around a cylindrical space of radius R and the other around a cylindrical space of radius $1/R$, then the winding number of one theory seems to match the momentum number (momentum, like about everything else, is quantized) of the other theory.

In other words, T-duality can relate a string theory with a large compactified radius to a different string theory with a small compactified radius (or, alternately, wide cylinders with narrow cylinders). Specifically, for closed strings, T-duality relates the following types of string theories:

- ✔ Type IIA and Type IIB superstring theories
- ✔ Type HO and Type HE superstring theories

The case for open strings is a bit less clear. When a dimension of superstring space-time is compactified into a circle, an open string doesn't wind around that dimension, so its winding number is 0. This means that it corresponds to a string with momentum 0 — a stationary string — in the dual superstring theory.

The end result of T-duality is an implication that Type IIA and IIB superstring theories are really two manifestations of the same theory, and Type HO and HE superstring theories are really two manifestations of the same theory.

Strong-weak duality: S-duality

Another duality that was known in 1995 is called *S-duality,* which stands for *strong-weak duality.* The duality is connected to the concept of a *coupling constant,* which is the value that tells the interaction strength of the string by describing how probable it is that the string will break apart or join with other strings.

The coupling constant, g, in string theory describes the interaction strength due to a quantity known as the *dilation field,* ϕ. If you had a high positive dilation field ϕ, the coupling constant $g = e\phi$ becomes very large (or the theory becomes strongly coupled). If you instead had a dilation field $-\phi$, the coupling constant $g = e^-\phi$ becomes very small (or the theory becomes weakly coupled).

Because of the mathematical methods (see nearby sidebar "Perturbation theory: String theory's method of approximation") that string theorists have to use to approximate the solutions to string theory problems, it was very hard to determine what would happen to string theories that were strongly coupled.

In S-duality, a strong coupling in one theory relates to a weak coupling in another theory, in certain conditions. In one theory, the strings break apart and join other strings easily, while in the other theory they hardly ever do so. In the theory where the strings break and join easily, you end up with a chaotic sea of strings constantly interacting.

Trying to follow the behavior of individual strings is similar to trying to follow the behavior of individual water molecules in the ocean — you just can't do it. So what do you do instead? You look at the big picture. Instead of looking at the smallest particles, you average them out and look at the unbroken surface of the ocean, which, in this analogy, is the same as looking at the strong strings that virtually never break.

S-duality introduces Type I string theory to the set of dual theories that T-duality started. Specifically, it shows that the following dualities are related to each other:

✔ Type I and Type HO superstring theories

✔ Type IIB is S-dual to itself

If you have a Type I superstring theory with a very strong coupling constant, it's theoretically identical to a Type HO superstring theory with a very weak coupling constant. So these two types of theories, under these conditions, yield the exact same predictions for masses and charges.

Perturbation theory: String theory's method of approximation

The equations of string theory are incredibly complex, so they often can only be solved through a mathematical method of approximation called *perturbation theory*. This method is used in quantum mechanics and quantum field theory all the time and is a well-established mathematical process.

In this method, physicists arrive at a first-order approximation, which is then expanded with other terms that refine the approximation. The goal is that the subsequent terms will become so small so quickly that they'll cease to matter. Adding even an infinite number of terms will result in converging onto a given value. In mathematical speak, *converging* means that you keep getting closer to the number without ever passing it.

Consider the following example of convergence: If you add a series of fractions, starting with ½ and doubling the denominator each time, and you added them all together (½ + ¼ ⅛ + . . . well, you get the idea), you'll always get closer to a

value of 1, but you'll never quite reach 1. The reason for this is that the numbers in the series get small very quickly and stay so small that you're always just a little bit short of reaching 1.

However, if you add numbers that double (2 + 4 + 8 + . . . well, you get the idea), the series doesn't converge at all. The solution keeps getting bigger as you add more terms. In this situation, the solution is said to *diverge* or become infinite.

The dual resonance model that Veneziano originally proposed — and which sparked all of string theory — was found to be only a first-order approximation of what later came to be known as string theory. Work over the last 40 years has largely been focused on trying to find situations in which the theory built around this original first-order approximation can be absolutely proved to be finite (or convergent), and which also matches the physical details observed in our own universe.

Using two dualities to unite five superstring theories

Both T-duality and S-duality relate different string theories together. Here's a review of the existing string theory relationships:

- ✔ Type I and Type HO superstring theories are related by S-duality.
- ✔ Type HO and Type HE superstring theories are related by T-duality.
- ✔ Type IIA and Type IIB superstring theories are related by T-duality.

With these dualities (and other, more subtle ones, which relate IIA and IIB together with the heterotic string theories), relationships exist to transform one version of string theory into another one — at least for certain specially selected string theory conditions.

To solve these equations of duality, certain assumptions have to be made, and not all of them are necessarily valid in a string theory that would describe our own universe. For example, the theories can only be proved in cases of perfect supersymmetry, while our own universe exhibits (at best) broken supersymmetry.

String theory skeptics aren't convinced that these dualities in some specific states of the theories relate to a more fundamental duality of the theories at all levels. Physicist (and string theory skeptic) Lee Smolin calls this the pessimistic view, while calling the string theory belief in the fundamental nature of these dualities the optimistic view.

Still, in 1995 it was hard not to be in the optimistic camp (and, in fact, many had never stopped being optimistic about string theory). The very fact that these dualities existed at all was startling to string theorists. It wasn't planned, but came out of the mathematical analysis of the theory. This was seen as powerful evidence that string theory was on the right track. Instead of falling apart into a bunch of different theories, superstring theory was actually pulling back together into one single theory — Edward Witten's M-theory — which manifested itself in a variety of ways.

The second superstring revolution begins: Connecting to the 11-dimensional theory

The period immediately following the proposal of M-theory has been called the "second superstring revolution," because it once again inspired a flurry of research into superstring theory. The research this time focused on understanding the connections between the existing superstring theories and between the 11-dimensional theory that Witten had proposed.

Witten wasn't the first one to propose this sort of a connection. The idea of uniting the different string theories into one by adding an 11th dimension had been proposed by Mike Duff of Texas A&M University, but it never caught on among string theorists. Witten's work on the subject, however, resulted in a picture where the extra dimension could emerge from the unifications inherent in M-Theory — one that prompted the string theory community to look at it more seriously.

In 1994, Witten and colleague Paul Townsend had discovered a duality between the 10-dimensional superstring theory and an 11-dimensional theory, which had been proposed back in the 1970s: supergravity.

Supergravity resulted when you took the equations of general relativity and applied supersymmetry to them. In other words, you introduced a particle called the gravitino — the superpartner to the graviton — to the theory. In the 1970s this was pretty much the dominant approach to trying to get a theory of quantum gravity.

What Witten and Townsend did in 1994 was take the 11-dimensional super-gravity theory from the 1970s and curl up one of the dimensions. They then showed that a membrane in 11 dimensions that has one dimension curled up behaves like a string in 10 dimensions.

Again, this is a recurrence of the old Kaluza-Klein idea, which comes up again and again in the history of string theory. By taking Kaluza's idea of adding an extra dimension (and Klein's idea of rolling it up very small), Witten showed that it was possible — assuming certain symmetry conditions — to show that dualities existed between the existing string theories.

There were still issues with an 11-dimensional universe. Physicists had shown supergravity didn't work because it allowed infinities. In fact, every theory except string theory allowed infinities. Witten, however, wasn't concerned about this because supergravity was only an approximation of M-theory, and M-theory would, by necessity, have to be finite.

It's important to realize that neither Witten nor anyone else proved that all five string theories could be transformed into each other in our universe. In fact, Witten didn't even propose what M-theory actually was.

What Witten did in 1995 was provide a theoretical argument to support the idea that there *could be* a theory — which he called M-theory — that united the existing string theories. Each known string theory was just an approximation of this hypothetical M-theory, which was not yet known. At low energy levels, he also believed that M-theory was approximated by the 11-dimensional supergravity theory.

Branes: Stretching Out a String

In a sense, the introduction of M-theory marks the end of "string theory," because it ceases to be a theory that contains only fundamental strings. M-theory also contains multidimensional membranes, called *branes*. Strings are only 1-dimensional objects, and therefore only one of the types of fundamental objects that make up the universe, according to the new M-Theory.

Branes have at least three key traits:

- ✔ Branes exist in a certain number of dimensions, from zero to nine.
- ✔ Branes can contain an electrical charge.
- ✔ Branes have a tension, indicating how resistant they are to influence or interaction.

String theory became more complex with the introduction of multidimensional branes. The first branes, called *D-branes,* entered string theory in 1989. Another type of brane, called a *p-brane* (not to be confused with the term you used to tease your younger sibling with), was later introduced. Later work showed that these two types of branes were in fact the same thing.

Branes are objects of multiple dimensions that exist within the full 10-dimensional space required by string theory. In the language of string theorists, this full space is called the *bulk*.

One major reason that string theorists didn't originally embrace branes was because introducing more elaborate physical objects went against the goal of string theory. Instead of simplifying the theory and making it more fundamental, branes made the theory more complicated and introduced more types of objects that didn't appear to be necessary. These were the exact features of the Standard Model that string theorists hoped to avoid.

In 1995, though, Joe Polchinski proved that it wasn't possible to avoid them. Any consistent version of M-theory had to include higher-dimensional branes.

The discovery of D-branes: Giving open strings something to hold on to

The motivation for D-branes came from work by Joe Polchinski, Jin Dai, and Rob Leigh of the University of Texas, and independent work performed at the same time by Czech physicist Petr Hořava. While analyzing the equations of string theory, these physicists realized that the ends of open strings didn't just hover out in empty space. Instead, it was as if the end of the open string was attached to an object, but string theory at the time didn't have objects (other than strings) for it to attach to.

To solve this problem, the physicists introduced the *D-brane,* a surface that exists within the 10-dimensional superstring theory so open strings can attach to them. These branes, and the strings attached to them, are shown in Figure 11-1. (The "D" in D-brane comes from Johann Peter Gustav Lejeune Dirichlet, a German mathematician whose relationship to the D-brane comes from a special type of boundary condition, called the *Dirichlet boundary condition,* which the D-branes exhibit.)

Figure 11-1:
Open strings attach to the brane at each end. The ends can attach to the same brane or to different branes.

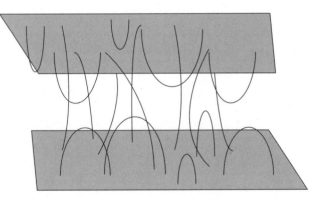

It's easiest to visualize these branes as flat planes, but the D-branes can exist in any number of dimensions from zero to nine, depending on the theory. A 5-dimensional D-brane would be called a D5-brane.

It's easy to see how quickly these D-branes can multiply. You could have a D5-brane intersecting a D3-brane, which has a D1-brane extending off of it. Open superstrings could have one end on the D1-brane and the other end on the D5-brane, or on some other D5-brane in another position, and D9-branes (extended in all nine dimensions of space-time) could be in the background of all of them. At this point, it's clear that it begins to be quite difficult to picture this 10-dimensional space or keep all the possible configurations straight in any meaningful way.

In addition, the D-branes can be either finite or infinite in size. Scientists honestly don't know the real limitations of how these branes behave. Prior to 1995, few people paid much attention to them.

Creating particles from p-branes

In the mid-1990s, Andrew Strominger performed work on another type of brane, called *p-branes,* which were solutions to Einstein's general relativity field equations. The *p* represents the number of dimensions, which again can go from zero to nine. (A 4-dimensional *p*-brane is called a 4-brane.)

The *p*-branes expanded infinitely far in certain directions but finitely far in others. In those finite dimensions, they actually seemed to trap anything that came near them, similar to the gravitational influence of a black hole. This work has provided one of the most amazing results of string theory — a way to describe some aspects of a black hole (see the section "Using branes to explain black holes").

In addition, the *p*-branes solved one problem in string theory: Not all of the existing particles could be explained in terms of string interactions. With the *p*-branes, Strominger showed that it was possible to create new particles without the use of strings.

A *p*-brane can make a particle by wrapping tightly around a very small, curled-up region of space. Strominger showed that if you take this to the extreme — picture a region of space that's curled up as small as possible — the wrapped *p*-brane becomes a massless particle.

According to Strominger's research with *p*-branes, not all particles in string theory are created by strings. Sometimes, *p*-branes can create particles as well. This is important because strings alone did not account for all the known particles.

Deducing that branes are required by M-theory

Strongly motivated by Edward Witten's proposal of M-theory, Joe Polchinski began working intently on D-branes. His work proved that D-branes weren't just a hypothetical construct allowed by string theory, but they were essential to any version of M-theory. Furthermore, he proved that the D-branes and *p*-branes were describing the same objects.

In a flurry of activity that would characterize the second superstring revolution, Polchinski showed that the dualities needed for M-theory only worked consistently in cases where the theory also contained higher dimensional objects. An M-theory that contained *only* 1-dimensional strings would be an inconsistent M-theory.

Polchinski defined what types of D-branes string theory allows and some of their properties. Polchinski's D-branes carried charge, which meant that they interacted with each other through something similar to the electromagnetic force.

A second property of D-branes is tension. The tension in the D-brane indicates how easily an interaction influences the D-brane, like ripples moving across a pool of water. A low tension means a slight disturbance results in large effects on the D-brane. A high tension means that it's harder to influence (or change the shape of) the D-brane.

If a D-brane had a tension of zero, then a minor interaction would have a major result — like someone blowing on the surface of the ocean and parting it like the Red Sea in *The Ten Commandments*. An infinite tension would mean the exact opposite: No amount of work would cause changes to the D-brane.

If you picture a D-brane as the surface of a trampoline, you can more easily visualize the situation. When the weight of your body lands on a trampoline, the tension in the trampoline is weak enough that it gives a bit, but strong enough that it does eventually bounce back, hurling you into the air. If the tension in the trampoline surface were significantly weaker or stronger, a trampoline would be no fun whatsoever; you'd either sink until you hit the ground, or you'd hit a flat, immovable trampoline that doesn't sink (or bounce) at all.

Together, these two features of the D-branes — charge and tension — meant that they aren't just mathematical constructs, but are tangible objects in their own right. If M-theory is true, D-branes have the capacity to interact with other objects and move from place to place.

Uniting D-branes and p-branes into one type of brane

Though Polchinski was aware of Strominger's work on *p*-branes — they discussed their projects over lunch regularly — both scientists thought that the two types of branes were distinct. Part of Polchinski's 1995 work on branes included the realization that they were actually one and the same object. At energy levels where predictions from string theory and general relativity match up, the two are equivalent.

It might seem odd that this hadn't occurred to either of the men before 1995, but there was no reason to expect that the two types of branes would be related to each other. To a layman, they sound basically the same — multidimensional surfaces existing in a 10-dimensional space-time. Why *wouldn't* you at least consider that they're the same things?

Well, part of the reason may be based on the specific nature of scientific research. When you're working in a scientific field, you are quite specific about the questions you're asking and the ways in which you're asking them. Polchinski and Strominger were asking different questions in different ways, so it never occurred to either of them that the answers to their questions might be the same. Their knowledge blinded them from seeing the commonalities. This sort of tunnel vision is fairly common and part of the reason why sharing research is so encouraged within the scientific community.

Similarly, for a laymen, the dramatic differences between these two types of branes are less clear. Just as someone who doesn't study much religion may be confused by the difference between Episcopalian and Catholic theological doctrines, to a priest of either religion the differences are well-known, and the two are seen as extremely distinct.

In the case of branes, though, the laymen would have had clearer insight on the issue than either of the experts. The very details that made D-branes and p-branes so intriguing to Polchinski and Strominger hindered their ability to see past the details to the commonalities — at least until 1995, when Polchinski finally saw the connection.

Because of equivalence, both D-branes and p-branes are typically just referred to as branes. When referencing their dimensionality, the p-brane notation is usually the one used. Some physicists still use the D-brane notation because there are other types of branes that physicists talk about. (For the remainder of this book, I mainly refer to them as branes, thus saving wear and tear on my keyboard's D key.)

Using branes to explain black holes

One of the major theoretical insights that string theory has offered is the ability to understand some black hole physics. These are directly related to work on p-branes, which, in certain configurations, can act something like black holes.

The connection between branes and black holes was discovered by Andrew Strominger and Cumrun Vafa in 1996. This is one of the few aspects of string theory that can be cited as actively confirming the theory in a testable way, so it's rather important.

The starting point is similar to Strominger's work on p-branes to create particles: Consider a tightly curled region of a space-dimension that has a brane wrapped around it. In this case, though, you're considering a situation in which gravity doesn't exist, which means you can wrap multiple branes around the space.

The brane's mass limits the amount of electromagnetic charge the brane can contain. A similar phenomenon happens with electromagnetically charged black holes. These charges create an energy density, which contributes to the mass of the black hole. This places a limit on the amount of electromagnetic charge a stable black hole can contain.

In the case where the brane has the maximum amount of charge — called an *extremal configuration* — and the case where the black hole has the maximum amount of charge — called an *extremal black hole* — the two cases share some properties. This allows scientists to use a thermodynamic model of an extremal configuration brane wrapped around extra dimensions to extract the thermodynamic properties that scientists would expect to obtain from an extremal black hole. Also, you can use these models to relate near-extremal configurations with near-extremal black holes.

Black holes are one of the mysteries of the universe that physicists would most like to have a clear explanation for. For more details on how string theory relates to black holes, skip ahead to Chapter 14.

String theory wasn't built with the intention of designing this relationship between wrapped branes and black holes. The fact that an artifact extracted purely from the mathematics of string theory would correlate so precisely with a known scientific object like a black hole, and one that scientists specifically want to study in new ways, was seen by everyone as a major step in support of string theory. It's just too perfect, many think, to be mere coincidence.

Getting stuck on a brane: Brane worlds

With the introduction of all of these new objects, string theorists have begun exploring what they mean. One major step is the introduction of *brane world* scenarios, where our 3-dimensional universe is actually a 3-brane.

Ever since the inception of string theory, one of the major conceptual hurdles has been the addition of extra dimensions. These extra dimensions are required so the theory is consistent, but we certainly don't seem to experience more than three space dimensions. The typical explanation has been to compactify the extra six dimensions into a tightly wound object roughly the size of the Planck length.

In the brane world scenarios, the reason we perceive only three spatial dimensions is that we live inside a 3-brane. There's a fundamental difference between the space dimensions on the brane and those off the brane.

The brane world scenarios are a fascinating addition to the possibilities of string theory, in part because they may offer some ways in which we can have consistent string theories without resorting to elaborate compactification scenarios. Not everyone is convinced, however, that compactifications can be eliminated from the theory and even some brane world theories include compactification as well.

In the "Infinite dimensions: Randall-Sundrum models" section later in this chapter, I look at some specific brane world scenarios that have been proposed, which offer some intriguing explanations for aspects of our universe, such as how to resolve the hierarchy problem (from Chapter 8). In Chapter 15, the idea of brane worlds allows you to consider the possibility of escaping our universe and traveling to a different universe on another brane!

Matrix Theory as a Potential M-Theory

A year after the proposal of M-theory, Leonard Susskind introduced a suggestion for what the "M" could stand for. *Matrix theory* proposes that the fundamental units of the universe are 0-dimensional point particles, which Susskind calls *partons* (or D0-branes). (No, these particles have nothing to do with the buxom Miss Dolly Parton.) These partons can assemble into all kinds of objects, creating the strings and branes required for M-theory. In fact, most string theorists believe that matrix theory is equivalent to M-theory.

Matrix theory was developed by Leonard Susskind, Tom Banks, Willy Fischler, and Steve Shenker in the year after Witten proposed M-theory. (The paper on the topic wasn't published until 1997, but Susskind presented the concept at a 1996 string theory conference prior to publication.) The theory is also approximated by 11-dimensional supergravity, which is one of the reasons string theorists think it's appropriate to consider it equal to M-theory.

The name "parton," which Susskind uses in his book *The Cosmic Landscape* (and I've used here) to describe these 0-dimensional branes, comes from a term used by the Nobel Prize-winning quantum physicist (and string theory skeptic) Richard P. Feynman. Both Feynman and his colleague and rival Murray Gell-Mann were working to figure out what made up hadrons. Though Gell-Mann proposed the quark model, Feynman had described a more vague theory where hadrons were made up of smaller pieces that he just called partons.

One intriguing aspect of the partons, noted by Witten, is that as they get close to each other, it becomes impossible to tell where the partons actually are. This may be reminiscent of the uncertainty principle in quantum mechanics, in which the position of a particle can't be determined with absolute precision, even mathematically (let alone experimentally). It's impossible to test this the same way scientists can test the uncertainty principle, because there's no way to isolate and observe an individual parton. Even light itself would be made up of a vast number of partons, so "looking" at a parton is impossible.

Unfortunately, the mathematics involved in analyzing matrix theory is difficult, even by the standards string theorists use. For now, research continues, and string theorists are hopeful that new insights may show more clearly how matrix theory can help shed light on the underlying structure of M-theory.

Gaining Insight from the Holographic Principle

Another key insight into string theory comes from the *holographic principle*, which relates a theory in space to a theory defined only on the boundary of that space. The holographic principle isn't strictly an aspect of string theory (or M-theory), but applies more generally to theories about gravity in any sort of space. Because string theory is such a theory, some physicists believe the holographic principle will lie at the heart of it.

Capturing multidimensional information on a flat surface

It turns out, as shown by Gerard 't Hooft in 1993 (and developed with much help from Leonard Susskind), the amount of "information" a space contains may be related to the area of a region's boundary, not its volume. (In quantum field theory, everything can be viewed as information.) In short, the holographic principle amounts to the following two postulates:

✔ A gravitational theory describing a region of space is equivalent to a theory defined only on the surface area that encloses the region.

✔ The boundary of a region of space contains at most one piece of information per square Planck length.

In other words, the holographic principle says that everything that happens in a space can be explained in terms of information that's somehow stored on the surface of that space. For example, picture a 3-dimensional space that resides inside the 2-dimensional curled surface of a cylinder, as in Figure 11-2. You reside inside this space, but perhaps some sort of shadow or reflection resides on the surface.

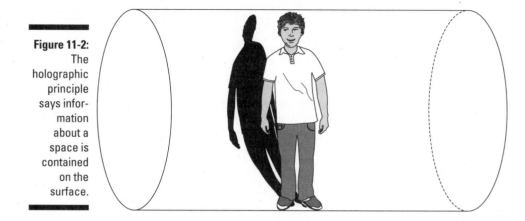

Figure 11-2:
The holographic principle says information about a space is contained on the surface.

Now, here's a key aspect of this situation that's missing from our example: A shadow contains only your outline, but in 't Hooft's holographic principle, *all* of the information is retained. (See the nearby sidebar, "Inside a hologram.")

Another example, and one that is perhaps clearer, is to picture yourself inside a large cube. Each wall of the cube is a giant television screen, which contains images of the objects inside the cube. You could use the information contained on the 2-dimensional surface of the space to reconstruct the objects within the space.

Again, though, this example falls short because not all of the information is encoded. If I were to have objects blocking me in all six directions, my image wouldn't be on any of the screens. But in the holographic principle view of the universe, the information on the surface contains all the information that exists within the space.

Connecting the holographic principle to our reality

The holographic principle is totally unexpected. You'd think that the information needed to describe a space would be proportional to the volume of that space. (Note that in the case of more than three space dimensions, "volume" isn't a precise term. A 4-dimensional "hypervolume" would be length times width times height times some other space direction. For now, you can ignore the time dimension.)

Inside a hologram

A *hologram* is a 2-dimensional image that contains all the 3-dimensional information of an object. When viewing a hologram, you can tilt the image and see the orientation of the shape move. It's as if you see the object in the picture from a different angle. The process of making a hologram is called *holography*.

This is achieved through the interference patterns in light waves. The process involves using a laser — so all of the light has exactly the same wavelength — and reflecting it off of the object onto a film. (When I performed this experiment in my college Optics class, I used a small plastic horse.)

As the light strikes the film, it records interference patterns that, when properly developed, allow the film to encode the information about the 3-dimensional shape that was holographed. The encoded information then has to be decoded, which means the laser light again has to be shown through the film in order to see the image.

"White light" holograms exist, which don't need laser light to view them. These are the holograms that you're most familiar with, which manifest their image in ordinary light.

You can consider this principle in two ways:

- ✔ Our universe is a 4-dimensional space that is equivalent to some 3-dimensional boundary.
- ✔ Our universe is a 4-dimensional boundary of a 5-dimensional space, which contains the same information.

In scenario 1, we live in the space inside the boundary, and in scenario 2, we are on the boundary, reflecting a higher order of reality that we don't perceive directly. Both theories have profound implications about the nature of the universe we live in.

Considering AdS/CFT correspondence

Though presented in 1993, even Leonard Susskind says he thought it would be decades before there would be any way to confirm the holographic principle. Then, in 1997, Argentinean physicist Juan Maldacena published a paper, inspired by the holographic principle, that proposed something called the *anti-de Sitter/conformal field theory correspondence,* or *AdS/CFT correspondence,* which brought the holographic principle to center stage in string theory.

In Maldacena's AdS/CFT correspondence, he proposed a new duality between a gauge theory defined on a 4-dimensional boundary (three space dimensions and one time dimension) and a 5-dimensional region (four space dimensions and one time dimension). In essence, he showed that there are circumstances in which the holographic principle scenario 2 is possible (see the preceding section).

As usual in string theory, one of those conditions is unbroken supersymmetry. In fact, the theoretical world he studied had the most amount of supersymmetry possible — it was maximally supersymmetrical.

Another condition was that the 5-dimensional region was something called an *anti-de Sitter space,* which means it had negative curvature. Our universe (at least at present) is more similar to a de Sitter space, as mentioned in Chapter 9. As such, it hasn't yet been proved that the AdS/CFT correspondence (or something similar) specifically applies to our own universe (though thousands of papers have been written on the subject).

Even if the duality turns out not to be completely true, a growing body of theoretical work supports the idea that there is some sort of correspondence between string theory and gauge theory, even if only at some low levels of approximation. Calculations that are hard in one version of the theory may actually be easy in the other one, meaning that it may be crucial in figuring out how to complete the theory. This has helped support the idea that the holographic principle may ultimately prove to be one of the fundamental principles of M-theory.

The holographic principle, and specifically the AdS/CFT correspondence, may also help scientists further understand the nature of black holes. The entropy (or disorder) of a black hole is proportional to the surface area of the black hole, not its volume. This is one of the arguments in support of the holographic principle, because it's believed that it would offer further physical explanation of black holes.

String Theory Gets Surprised by Dark Energy

The discovery of dark energy in 1998 meant that our universe needed to have a positive cosmological constant. The problem is that all of the string theories were built in universes with negative cosmological constants (or a zero value). When work did discover possible ways to incorporate a positive cosmological constant, it resulted in a theory that has a vast number of possibilities!

Dark energy is an energy that seems to fill much of the universe and causes space-time to expand. By current estimates, more than 70 percent of the universe is comprised of dark energy.

Prior to the 1998 discovery, the assumption was that the universe had a zero cosmological constant, so all the work done in string theory was focused on that sort of a universe. With the discovery of dark matter, priorities had to change. The search was on for a universe that had a positive cosmological constant.

Joe Polchinski and Raphael Bousso extended others' earlier research by experimenting with extra dimensions that had *electric flux* (a number that represents the intensity of an electric field through a surface) wrapped around them. Branes carried charge, so they could also have flux. This construction had the potential to limit some parameters of the theory in a way that couldn't vary continuously.

In 2003, a Stanford group including Renata Kallosh, Andrei Linde, Shamit Kachru, and Sandip Trivedi released a paper that showed ways to extend the Polchinski-Bousso thinking to construct string theories with a positive cosmological constant. The trick was to create a universe and then wrap it with branes and anti-branes to contain the electric and magnetic flux. This introduced the potential for two effects:

✔ It allowed a small positive cosmological constant.

✔ It stabilized the extra dimensions in string theory.

On the surface, this would seem to be an excellent outcome, providing two necessary components to string theory. Unfortunately, there was one little problem — there were far too many solutions!

Considering Proposals for Why Dimensions Sometimes Uncurl

Most string theory proposals have been based on the concept that the extra dimensions required by the theory are curled up so small that they can't be observed. With M-theory and brane worlds, it may be possible to overcome this restriction.

A few scenarios have been proposed to try to describe a mathematically coherent version of M-theory, which would allow the extra dimensions to be extended. If any of these scenarios hold true, they have profound implications for how (and where) physicists should be looking for the extra dimensions of string theory.

Measurable dimensions

One model that has gotten quite a bit of attention was proposed in 1998 by Savas Dimopoulos, Nima Arkani-Hamed, and Gia Dvali. In this theory, some of the extra dimensions could be as large as a millimeter without contradicting known experiments, which means that it may be possible to observe their effects in experiments conducted at CERN's Large Hadron Collider (LHC). (This proposal has no unique name, but I call it *MDM* for *millimeter dimension model.* Who knows, maybe it'll catch on!)

When Dimopoulos introduced MDM at a 1998 supersymmetry conference, it was actually something of a subversive act. He was making a bold statement: Extra dimensions were as important, if not more so, than supersymmetry.

Many physicists believe that supersymmetry is the key physical principle that will prove to be the foundation of M-theory. Dimopoulos proposed that the extra dimensions — previously viewed as an unfortunate mathematical complication to be ignored as much as possible — could be the fundamental physical principle M-theory was looking for.

In MDM, a pair of extra dimensions could extend as far as a millimeter away from the 3-dimensional brane that we reside on. If they extend much more than a millimeter, someone would have noticed by now, but at a millimeter, the deviation from Newton's law of gravity would be so slight that no one would be any the wiser. So because gravity is radiating out into extra dimensions, it would explain why gravity is so much weaker than the brane-bound forces.

The way this works is everything in our universe is trapped on our 3-dimensional brane *except gravity,* which can extend off of our brane to affect the other dimensions. Unlike in string theory, the extra dimensions wouldn't be noticeable in experiments except for gravity probes, and in 1998, gravity hadn't been tested at distances shorter than a millimeter.

Now, don't get too excited yet. Experiments have been done to look for these extra millimeter-sized dimensions and, it turns out, they probably don't exist. Experiments show that the dimensions have to be at least as small as a tenth of a millimeter, but that's still far larger than in most other string theory scenarios. Instead of requiring the 10^{19} GeV (giga-electronvolts, a unit of energy) needed to explore the Planck length, exploring a millimeter would require only 1,000 GeV — still within the range of CERN's LHC!

Infinite dimensions: Randall-Sundrum models

If a millimeter-sized dimension turned heads, the 1999 proposal by Lisa Randall and Raman Sundrum was even more spectacular. In these *Randall-Sundrum models,* gravity behaves differently in different dimensions, depending on the geometry of the branes.

Yet another string theory: F-theory

Another theory that sometimes gets discussed is called F-theory (the name is a joking reference to the idea that the M in M-theory stands for mother). Cumrun Vafa proposed F-theory in 1996 after noticing that certain complicated solutions of Type IIB string theory could be described in terms of a simpler solution of a different theory with 12 dimensions, up from the 10 dimensions of superstrings or the 11 dimensions of M-theory. Unlike M-theory, where all the dimensions of space-time are treated on equal footing, two of the dimensions of F-theory are fundamentally different than the rest: They *always* have to be curled up. So now to get to three space dimensions, we have eight small dimensions instead of six!

This makes it seem as though the theory is getting more complicated, but in fact the F-theory description is often simpler. These eight dimensions include not only all the information from the previous six, but also information about what branes exist in the solution (those setups could get complicated). This is an example of a common theme in the development of string theory; more and more of the theory's details, such as what particles exist and how they interact, or what branes live where, can be described simply in terms of the geometry of the extra dimensions. This geometry is often easier to understand and analyze.

F-theory has been receiving more attention in the past few years because its rich structure allows solutions that reproduce many of the phenomena of the Standard Model and GUT theories (see Chapter 12 for more on those).

In the original Randall-Sundrum model, called *RS1,* they propose a brane that sets the strength of gravity. In this *gravitybrane,* the strength of gravity is extremely large. As you move in a fifth dimension away from the gravitybrane, the strength of gravity drops exponentially.

An important aspect of the RS1 model is that the strength of gravity depends only on the position within the fifth dimension. Because our entire 3-brane (this is a brane world scenario, where we're trapped on a 3-brane of space) is at the same fifth-dimensional position, gravity is consistent everywhere in the 3-brane.

In a second scenario, called *RS2,* Randall and Sundrum realized that the 3-brane that we're stuck in could have its own gravitational influence. Though gravitons can drift away from the 3-brane into other dimensions, they can't get very far because of the pull of our 3-brane. Even with large dimensions, the effects of gravity leaking into other dimensions would be incredibly small. Randall and Sundrum called the RS2 model *localized gravity.*

In both of these models, the key feature is that gravity on our own 3-brane is essentially always the same. If this weren't the case, we'd have noticed the extra dimensions before now.

In 2000, Lisa Randall proposed another model with Andreas Karch called *locally localized gravity.* In this model, the extra dimension contained some negative vacuum energy. It goes beyond the earlier models, because it allows gravity to be localized in different ways in different regions. Our local area looks 4-dimensional and has 4-dimensional gravity, but other regions of the universe might follow different laws.

Understanding the Current Landscape: A Multitude of Theories

As far back as 1986, Andrew Strominger found that there was a vast number of consistent string theory solutions and observed that all predictive power may have been lost. Actually, when considering a negative cosmological constant (or zero), you apparently end up with an infinite number of possible theories.

With a positive cosmological constant — as needed in our universe, thanks to dark energy — things get better, but not by much. There are now a finite number of ways to roll up the branes and anti-branes so as to obtain a positive cosmological constant. How many ways? Some estimates have indicated as many as 10^{500} possible ways to construct such a string theory!

This is an enormous problem if the goal of string theory was to develop a single unified theory. The vision of both the first and second superstring revolutions (or at least the vision guiding some bandwagoners who jumped on board) was a theory that would describe our universe with no experimental observations required.

In 2003, Leonard Susskind published "The Anthropic Landscape of String Theory," in which he very publicly gave up the idea that a unique string theory would be discovered. In the paper, Susskind introduced the concept of "the landscape" of string theories: a vast number of mathematically consistent possible universes, some of which actually exist. Susskind's *string theory landscape* was his solution to the unfathomable number of possible string theories.

But with so many possibilities, does the theory have any predictive power? Can we use a theory if we don't know what the theory is?

The anthropic principle requires observers

Susskind's proposed solution involves relying on something known as the *anthropic principle*. This principle indicates that the reason the universe has the properties it does is because we're here to observe them. If it had vastly different properties, we wouldn't exist. Other areas of the *multiverse* may have different properties, but they're too far away for us to see.

REMEMBER

The anthropic principle was coined by Cambridge astrophysicist Brandon Carter in 1974. It exists in two basic versions:

- ✔ **Weak anthropic principle:** Our location (or region) of space-time possesses laws such that we exist in it as observers.

- ✔ **Strong anthropic principle:** The universe is such that there has to exist a region of space-time within it that allows observers.

If you're reading these two variations of the anthropic principle and scratching your head, you're in good company. Even string theorists who are now embracing the anthropic principle — such as Susskind and Joe Polchinski — once despised it as totally unscientific. This is in part due to the fact that the anthropic principle (in its strong form) is sometimes invoked to require a supernatural designer of the universe, something that most scientists (even religious ones) try to avoid in their scientific work. (Ironically, it is also often used, in the weak form, as an argument *against* a supernatural designer, as Susskind does in his book *The Cosmic Landscape*.)

For the anthropic principle to make sense, you have to consider an array of possible universes. Figure 11-3 shows a picture of the energy levels of possible universes, where each valley represents a particular set of string theory parameters.

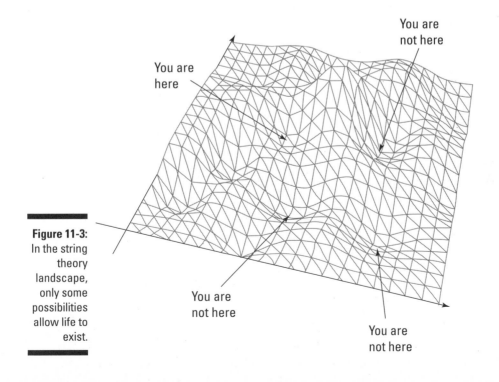

You are not here

You are here

Figure 11-3: In the string theory landscape, only some possibilities allow life to exist.

You are not here

You are not here

According to the weak anthropic principle, the only portions of the multiverse we can ever observe are the ones where these parameters allow us to exist.

In this sense, the weak anthropic principle is almost a given — it's just always going to be true. That's part of the point of it. Because we're here, we can use the fact that we're here to explain the properties the universe has. In the string theory landscape, so many possibilities are out there that ours is just one of them, which has happened to come into being, and we're lucky enough to be here.

If the string theory landscape represents all the universes that are possible, the multiverse represents all the universes that actually exist. Distant regions of the multiverse may have radically different physical properties than those that we observe in our own section.

This concept is similar to Lisa Randall's locally localized gravity (see the "Infinite dimensions: Randall-Sundrum models" section earlier in this chapter), where only our local region exhibits the gravity that we know and love in three space dimensions. Other regions could have five or six space dimensions, but that doesn't matter to us, because they're so far away that we can't see them. These other regions are different parts of the multiverse.

In 1987, Nobel Prize-winner Steven Weinberg added a bit of credibility to the field. Using reasoning based on the anthropic principle, he analyzed the cosmological constant required to create a universe like ours. His prediction was a very small positive cosmological constant, only about one order of magnitude off from the value found more than a decade later.

This is a frequently cited case of when the anthropic principle led to a testable prediction, but I've never been particularly convinced that it's that meaningful. Clearly, our universe is one in which galaxies formed the way they have — not too fast or too slow. Using that fact is totally uncontroversial as a means of determining the cosmological constant, but the anthropic principle goes further. It doesn't just determine the cosmological constant, it supposedly explains why the cosmological constant has that value.

The key feature of anthropic reasoning is that there exists an entire multiverse of possibilities. If there's just one universe, we have to explain *why* that universe is so perfectly suited for humans to exist. But if there are a vast number of universes, and they take on a wide range of parameters, then probability dictates that every once in a while a universe like ours will spring up, resulting in life forms and observers like us.

Disagreeing about the principle's value

Since its introduction in 1974, the anthropic principle has invoked passion among scientists. It's safe to say that most physicists don't consider invoking the anthropic principle to be the best scientific tactic. Many physicists see it as giving up on an explanation, and just saying "it is what it is."

At Stanford, Leonard Susskind and his colleagues seem to be embracing the anthropic principle. To hear (or read) Susskind on the subject, the string theory community is quickly jumping on board. It's unclear whether the movement is spreading quite as intently as this rhetoric implies, though.

One barometer could be the literature. Out of 13 string theory books (written after 2003 — 8 popular books, 5 textbooks) within my reach at this moment, here are the statistics:

- 5 make no mention of the anthropic principle in the index

- 2 discuss the anthropic principle for precisely one paragraph

- 2 contain more general discussions of the anthropic principle, lasting about two pages

- 2 attack the landscape and anthropic principle as major failures of the theory, devoting roughly an entire chapter to the concept

- 2 argue that the anthropic principle is crucial to understanding our universe (and one of those is written by Susskind himself)

On the other hand, a search of the arXiv.org theoretical physics database shows 218 hits on a search of the phrase "anthropic." Searching on "anthropic principle" obtains 104 hits, and adding words such as "string" and "brane" only causes it to drop from there. For comparison, searching on "string theory," "cosmological constant," or even the far less popular "loop quantum gravity" result in so many hits that the search cuts off at only 1,000 papers. So the jury is certainly still out on how well the string theory community has adopted the anthropic principle.

Some string theorists, such as David Gross, appear to be strongly opposed to anything that even hints at the anthropic principle. A large number of string theorists bought into it based on the idea — championed by Witten's promise of M-theory in 1995 — that there would be a single theory at the end of the rainbow.

String theorists seem to be turning to the anthropic principle mostly out of a lack of other options. This certainly seems to be the case for Edward Witten, who has made public statements indicating he might be unenthusiastically turning toward anthropic thinking.

We end the chapter in many ways worse off than we began. Instead of five distinct string theory solutions, we have 10^{500} or so. It's unclear what the fundamental physical properties of string theory are in a field of so many options. The only hope is that new observations or experiments will provide some sort of clue about which aspect of the string theory landscape to explore next.

Chapter 12

Putting String Theory to the Test

*N*o matter how impressive string theory is, without experimental confirmation, it's nothing but mathematical speculation. As discussed in Chapter 4, science is an interplay of theory and experiment. String theory attempts to structure the experimental evidence around a new theoretical framework.

One problem with string theory is that the energy required to get direct evidence for the distinct predictions of the theory is typically so high that it's very hard to reach. New experimental methods, such as the Large Hadron Collider (described later in this chapter), are expanding our ability to test in higher energy ranges, possibly leading to discoveries that more strongly support string theory predictions, such as extra dimensions and supersymmetry. Probing the strings themselves requires massive amounts of energy that are still far away from any experimental exploration.

In this chapter, my goal is to look at different ways that string theory can be tested, so it can be either verified or disproved. First, I explain the work that still needs to be done to complete the theory so it can make meaningful predictions. I also cover a number of experimental discoveries that would pose complications for string theory. Then I discuss ways of proving that our universe does contain supersymmetry, a key assumption required by string theory. Finally, I outline the testing apparatus — those created in deep space and particle accelerators created on Earth.

Understanding the Obstacles

As discussed in Chapter 11, string theory isn't complete. There are a vast number of different string theory solutions — literally billions of billions of billions of billions of different possible variants of string theory, depending on the parameters introduced into the theory. So, in order to test string theory, scientists have to figure out which predictions the theory actually makes.

Before testing on string theory can take place, physicists need to filter through the massive possible number of solutions to find a manageable amount that may describe our universe. Most of the current tests related to string theory are measurements that are helping to define the current parameters of the theory. Then, after the remaining theoretical solutions are somehow assessed in a reasonable way, scientists can begin testing the unique predictions they make.

There are two features common to (almost) all versions of string theory, and scientists who are looking for evidence of string theory are testing these ideas even now:

- ✔ Supersymmetry
- ✔ Extra dimensions

These are string theory's two cornerstone ideas (aside from the existence of strings themselves, of course), which have been around since the theory was reformulated into superstring theory in the 1970s. No theory that has tried to eliminate them has lasted very long.

Testing an incomplete theory with indistinct predictions

Right now, there is a great deal of confusion over what physical properties (other than supersymmetry and extra dimensions) lie at the heart of string theory. The holographic principle, anthropic principle, brane world scenarios, and other such approaches are becoming more popular, but scientists don't know for certain how they apply in the case of our universe.

The energy constraints on string theory experiments are obviously a big obstacle, but I think for most skeptical theorists, lack of specific, distinct experiments is the more disturbing issue. The variants of string theory make few distinct predictions, so it's hard to even think about testing it. Scientists can continue to test aspects of the Standard Model, to make sure that string theory predictions remain consistent, and they can look for properties such as supersymmetry or extra dimensions, but these are very general predictions, many of which are made not just by string theory. The first step in testing string theory is to figure out what the theory is telling us that is distinct from other theories.

Test versus proof

There's really no way to prove something like string theory, as a whole. You can prove that a specific prediction (such as supersymmetry, which I get to later in this chapter) is true, but that doesn't prove that the theory as a whole is true. In a very real sense, string theory can never be proved; it can just meet the test of time, the same way that other theories have done.

For scientists, this slight distinction is known and accepted, but there's some confusion about it among nonscientists. Most people believe that science proves things about the laws of nature beyond a shadow of a doubt, but the truth is that science dictates there is *always* a shadow of a doubt in any theory.

A theory can be tested in two ways. The first is to apply the theory to explain existing data (called a *postdiction*). The second is to apply the theory to determine new data, which experiments can then look for. String theory has been very successful at coming up with postdictions, but it hasn't been as successful at making clear predictions.

String theory, as Chapter 17 explains, has some valid criticisms that need to be addressed. Even if they are addressed, string theory will never be proved, but the longer it makes predictions that match experiments, the more support it will gain.

For this to happen, of course, string theory has to start making predictions that can be tested.

Testing Supersymmetry

One major prediction of string theory is that a fundamental symmetry exists between bosons and fermions, called supersymmetry. For each boson there exists a related fermion, and for each fermion there exists a related boson. (Bosons and fermions are types of particles with different spins; Chapter 8 has more detail about these particles.)

Finding the missing sparticles

Under supersymmetry, each particle has a superpartner. Every boson has a corresponding fermionic superpartner, just as every fermion has a bosonic superpartner. The naming convention is that fermionic superpartners end in "–ino," while bosonic superpartners start with an "s." Finding these superpartners is a major goal of modern high-energy physics.

The problem is that without a complete version of string theory, string theorists don't know what energy levels to look at. Scientists will have to keep exploring until they find superpartners and then work backward to construct a theory that contains the superpartners. This seems only slightly better than the Standard Model of particle physics, where the properties of all 18 fundamental particles have to be entered by hand.

Also, there doesn't appear to be any fundamental theoretical reason *why* scientists haven't found superpartners yet. If supersymmetry does unify the forces of physics and solve the hierarchy problem, then scientists would expect to find low-energy superpartners. (The search for the Higgs boson has undergone these same issues within the Standard Model framework for years. It has yet to be detected experimentally either.)

Instead, scientists have explored energy ranges into a few hundred GeV, but still haven't found any superpartners. So the lightest superpartner would appear to be heavier than the 17 observed fundamental particles. Some theoretical models predict that the superpartners could be 1,000 times heavier than protons, so their absence is understandable (heavier particles often tend to be more unstable and collapse into lower-energy particles if possible) but still frustrating.

Right now, the best candidate for a way to find supersymmetric particles outside of a high-energy particle accelerator (see the later section "Large Hadron Collider (LHC)") is the idea that the dark matter in our universe may actually be the missing superpartners (see the later section "Analyzing dark matter and dark energy").

Testing implications of supersymmetry

If supersymmetry exists, then some physical process takes place that causes the symmetry to become spontaneously broken as the universe goes from a dense high-energy state into its current low-energy state. In other words, as the universe cooled down, the superpartners had to somehow decay into the particles we observe today. If theorists can model this spontaneous symmetry-breaking process in a way that works, it may yield some testable predictions.

The main problem is something called the *flavor problem*. In the standard model, there are three flavors (or generations) of particles. Electrons, muons, and taus are three different flavors of leptons.

In the Standard Model, these particles don't directly interact with each other. (They can exchange a gauge boson, so there's an indirect interaction.) Physicists assign each particle numbers based on its flavor, and these numbers are a conserved quantity in quantum physics. The electron number, muon number, and tau numbers don't change, in total, during an interaction. An electron, for example, gets a positive electron number but gets 0 for both muon and tau numbers.

Because of this, a muon (which has a positive muon number but an electron number of zero) can never decay into an electron (with a positive electron number but a muon number of zero), or vice versa. In the Standard Model and in supersymmetry, these numbers are conserved, and interactions between the different flavors of particles are prohibited.

However, our universe doesn't have supersymmetry — it has *broken supersymmetry*. There is no guarantee that the broken supersymmetry will conserve the muon and electron number, and creating a theory of spontaneous supersymmetry breaking that keeps this conservation intact is actually very hard. Succeeding at it may provide a testable hypothesis, allowing for experimental support of string theory.

Testing Gravity from Extra Dimensions

The testing of gravity produces a number of ways to see if string theory predictions are true. When physicists test for gravity outside of our three dimensions, they

- ✔ Search for a violation of the inverse square law of gravity
- ✔ Search for certain signatures of gravity waves in the cosmic microwave background radiation (CMBR)

It may be possible that further research will result in other ways to determine the behavior of string theory or related concepts (see the nearby sidebar, "Detecting the holographic principle with gravity waves").

Testing the inverse square law

If extra dimensions are compactified in ways that string theorists have typically treated them, then there are implications for the behavior of gravity. Specifically, there might be a violation of the inverse square law of gravity, especially if gravitational force extends into these extra dimensions at small scales. Current experiments seek to test gravity to an unprecedented level, hoping to see these sorts of differences from the established law.

The behavior of gravity has been tested down to under a millimeter, so any compactified dimensions must be smaller than that. Recent models indicate that they may be as large as that, so scientists want to know if the law of gravitation breaks down around that level.

As of this book's publication, no evidence has been found to confirm the extra dimensions at this level, but only time will tell.

Searching for gravity waves in the CMBR

General relativity predicts that gravity moves in waves through space-time. Although string theory agrees with this prediction, in most string theory-based models of inflation, there are no observable gravity waves in the cosmic microwave background radiation (CMBR). Traditional inflation models that don't take string theory into account do predict CMBR gravity waves.

Again, this turns out to be a search for evidence against string theory, but this has a bit more weight behind it than some of the others. Although the string theory landscape has predictions for scenarios where relativity breaks down, there doesn't appear to be any mechanism in string theory for gravity waves in the CMBR, according to University of California cosmologist and string theorist Andrei Linde. (Linde made this statement in 2007 and work since then has produced some preliminary indications that string theory models of inflation may be compatible with gravity waves in the CMBR.)

At present, the evidence seems to be leaning toward there not being any gravity waves in the CMBR data. The Planck Surveyor spacecraft was successfully launched in May 2009, with even greater sensitivity than the current WMAP study. Scientists may get a more decisive take on whether these CMBR gravity waves exist at any time.

Detecting the holographic principle with gravity waves

Results from the GEO600 gravity wave detector in Germany may already have found evidence for the holographic principle, though the co-creator of the holographic principle is skeptical.

In 2007, Fermilab physicist Craig Hogan realized that if the bits of information on the surface of space are Planck length in size (as the holographic principle suggests), the bits of information contained inside the space have to be larger. He then predicted that this would cause some static in gravity wave detectors. And, sure enough, static was being detected by GEO600 in precisely the way predicted.

This would seem like an open and shut case, but there are many possible sources of this noise in the GEO600, and until they're eliminated everyone is cautiously optimistic. Plus, Hogan's paper is not so much a theory as a neat idea, and no one is exactly sure what it means — including string theory and holographic principle co-founder Leonard Susskind. Susskind told me in an e-mail that he doesn't understand how the holographic principle would result in gravitational wave noise.

Disproving String Theory Sounds Easier Than It Is

With any theory, it's typically easier to disprove it than to prove it, although one criticism of string theory is that it may have become so versatile that it can't be disproved. I elaborate on this concern in Chapter 17, but in the following sections I assume that string theorists can pull together a specific theory. Having a working theory in hand makes it easier to see how it could be proved wrong.

Violating relativity

String theories are constructed on a background of space-time coordinates, so physicists assume relativity is part of the environment. If relativity turns out to be in error, then physicists will need to revise this simplifying assumption, although it's unlikely that this alone would be enough to cause them to abandon string theory entirely (nor should it).

There are theories that predict errors in relativity, most notably the variable speed of light (VSL) cosmology theories of John Moffat, and Andreas Albrecht and João Magueijo. Moffat went on to create a more comprehensive revision of general relativity with his modified gravity (MOG) theories. These theories are addressed in Chapter 19, but they mean that the current assumptions of string theory contain errors.

Even in this case, though, string theory would survive. Elias Kiritsis and Stephon Alexander have both proposed VSL theories within the context of string theory. Alexander went on to do further work in this vein with the "bad boy of cosmology," João Magueijo, who is fairly critical of string theory as a whole.

Mathematical inconsistencies

Given that string theory exists only on paper right now, one major problem would be a definitive proof that the theory contained mathematical inconsistencies. This is the one area where string theory has proved most adaptable, successfully avoiding inconsistencies for more than 20 years.

Of course, scientists know that string theory isn't the whole story — the true theory is an 11-dimensional M-theory, which has not yet been defined. Work continues on various string theory approximations, but the fundamental theory — M-theory — may still prove to be nothing more than a myth (yet another word the M could stand for).

One weakness is in the attempt to prove string theory finite. In Chapter 17, you can read about the controversy over whether this has been achieved. (It appears that even among string theorists there's a growing acknowledgement that the theory hasn't been proved finite to the degree that it was once hoped it would be.)

To create his theory of gravity, Newton had to develop calculus. To develop general relativity, Einstein had to make use of differential geometry and develop (with the help of his friend Marcel Grossman) tensor calculus. Quantum physics was developed hand in hand with group representation theory by innovative mathematician Hermann Weyl. (*Group representation theory* is the mathematical study of how symmetries can act on vector spaces, which is at the heart of modern physics.)

Though string theory had already spawned innovative mathematics explorations, the fact that scientists don't have any complete version of M-theory implies to some that some key mathematical insight is missing — or that the theory simply doesn't exist.

Could Proton Decay Spell Disaster?

If one of the older attempts at unification of forces (called *grand unification theories* or GUTs) proves successful, it would have profound implications for string theory. One of the most elegant GUTs was the 1974 Georgi-Glashow model, proposed by Howard Georgi and Sheldon Glashow. This theory has one flaw: It predicts that protons decay, and experiments over the last 25 years have not shown this to be the case. Even if proton decay is detected, string theorists may be able to save their theory.

The Georgi-Glashow model allows quarks to transform into electrons and neutrinos. Because protons are made of specific configurations of quarks, if a quark inside a proton were to suddenly change into an electron, the proton itself would cease to exist as a proton. The nucleus would emit a new form of radiation as the proton decayed.

This quark transformation (and resulting proton decay) exists because the Georgi-Glashow model uses a SU(5) symmetry group. In this model, quarks, electrons, and neutrinos are the same fundamental kind of particle, manifesting in different forms. The nature of this symmetry is such that the particles can, in theory, transform from one type into another.

Of course, these decays can't happen very often, because we need protons to stick around if we're going to have a universe as we know it. The calculations showed that a proton decays at a very small rate: less than one proton every 10^{33} years.

This is a very small decay rate, but there's a way around it by having a lot of particles. Scientists created vast tanks filled with ultrapure water and shielded from cosmic rays that could interfere with protons (and give false decay readings). They then waited to see if any of the protons decayed.

After 25 years, there has been no evidence of proton decay, and these experiments are constructed so there could be as many as a few decays a year. The results from the Super-Kamiokande, a neutrino observatory in Japan, show that an average proton would take at least 10^{35} years to decay. To explain the lack of results, the Georgi-Glashow model has been modified to include longer decay rates, but most physicists don't expect to observe proton decay anytime soon (if at all).

If scientists did finally discover the decay of a proton, that would mean that the Georgi-Glashow model would need to be looked at anew. String theory gained success in part because of the failure of all other previous models, so if their predictions work, it may indicate poor prospects for string theory.

The string theory landscape remains as resilient as ever, and some predictions of string theory allow for versions that include proton decay. The decay timeframe predicted is roughly 10^{35} years — exactly the lower limit allowed by the Super-Kamiokande neutrino observatory.

The renewal of GUTs would not disprove string theory, even though the failure of GUT is part of the reason why string theory was originally adopted. String theories can now incorporate GUT in low-energy domains. But string theory can't tell us whether we should anticipate that GUT exists or protons decay. Maybe or maybe not, and string theory can deal with it either way. This is just one of the many cases where string theory shows a complete ambivalence to experimental evidence, which some critics say makes it "un-falsifiable" (as discussed at greater length in Chapter 17).

Looking for Evidence in the Cosmic Laboratory: Exploring the Universe

The problem with conducting experiments in string theory is that it requires massive amounts of energy to reach the level where the Standard Model and general relativity break down. Although I address manmade attempts to explore this realm in the next section, here I look at the different route the field of string cosmology takes — attempting to look into nature's own laboratory, the universe as a whole, to find the evidence that string theorists need to test their theories.

Using outer space rays to amplify small events

Among the various phenomena in the universe, two types produce large amounts of energy and may provide some insight into string theory: *gamma ray bursts (GRBs)* and *cosmic rays.*

Some physical events are hard to see because they

- Are very rare (like, possibly, proton decay)
- Are very small (like Planck-scale events or possible deviations in gravity's effects)
- Happen only at very high energies (like high-energy particle collisions)

Or, some combination of the three makes the event a challenge to witness. Scientists are unlikely to see these improbable events in laboratories on Earth, at least without a lot of work, so sometimes they look where they're more likely to find them. Because both GRBs and cosmic rays contain very high energies and take so long to reach us, scientists hope they can observe these hard-to-see events by studying the cosmic happenings.

For years, physicists had used this method to explore potential breakdowns in special relativity, but Italian physicist Giovanni Amelino-Camelia of the University of Rome realized in the mid-1990s that this process could be used to explore the Planck length (and energy) scale.

Gamma ray bursts

Exactly what causes a gamma ray burst is disputed, but it seems to happen when massive objects, such as a pair of neutron stars or a neutron star and a black hole (the most probable theories), collide with each other. These objects orbit around each other for billions of years, but finally collapse together, releasing energy in the most powerful events observed in the universe, depicted in Figure 12-1.

Figure 12-1:
When some stars die, they release massive bursts of energy.

Courtesy of NASA/Swift/Sonoma State University/A. Simonnet

The name *gamma ray bursts* clearly implies that most of this energy leaves the event in the form of gamma rays, but not all of it does. These objects release bursts of light across a range of different energies (or frequencies — energy and frequency of photons are related).

According to Einstein, all the photons from a single burst should arrive at the same time, because light (regardless of frequency or energy) travels at the same speed. By studying GRBs, it may be possible to tell if this is true.

Calculations based on Amelino-Camelia's work has shown that photons of different energy that have traveled for billions of years could, due to (estimated and possibly over-optimistic) quantum gravity effects at the Planck scale, have differences of about 1 one-thousandth of a second (0.001s).

The Fermi Gamma-ray Space Telescope (formerly the Gamma-ray Large Area Space Telescope, or GLAST) was launched in June 2008 as a joint venture between NASA, the U.S. Department of Energy, and French, German, Italian, Japanese, and Swedish government agencies. Fermi is a low-Earth orbit observatory with the precision required to detect differences this small.

So far, there's no evidence that Fermi has identified Planck scale breakdown of general relativity. To date it's identified a dozen gamma ray–only pulsars, a phenomenon that had never been observed before Fermi. (Prior to Fermi, *pulsars* — spinning and highly magnetized neutron stars that emit energy pulses — were believed to emit their energy primarily through radio waves.)

If Fermi (or some other means) does detect a Planck scale breakdown of relativity, then that will only increase the need for a successful theory of quantum gravity, because it will be the first experimental evidence that the theory does break down at these scales. String theorists would then be able to incorporate this knowledge into their theories and models, perhaps narrowing the string theory landscape to regions that are more feasible to work with.

Cosmic rays

Cosmic rays are produced when particles are sent out by astrophysical events to wander the universe alone, some traveling at close to the speed of light. Some stay bound within the galactic magnetic field, while others break free and travel between galaxies, traveling billions of years before colliding with another particle. These cosmic rays can be more powerful than our most advanced particle accelerators.

First of all, cosmic rays aren't really rays. They're stray particles in mostly three forms: 90 percent free protons, 9 percent alpha particles (two protons and two neutrons bound together — the nucleus of a helium atom), and 1 percent free electrons (beta minus particles, in physics-speak).

Astrophysical events — everything from solar flares to binary star collisions to supernovae — regularly spit particles out into the vacuum of space, so our planet (and, in turn, our bodies) are constantly bombarded with them. The particles may travel throughout the galaxy, bound by the magnetic field of the galaxy as a whole, until they collide with another particle. (Higher energy particles, of course, may even escape the galaxy.)

Fortunately for us, the atmosphere and magnetic field of Earth protect us from the most energetic of these particles so we aren't continuously dosed with intense (and lethal) radiation. The energetic particles are deflected or lose energy, sometimes colliding in the upper atmosphere to split apart into smaller, less energetic particles. By the time they get to us, we're struck with the less intense version of these rays and their offspring.

Cosmic rays have a long history as experimental surrogates. When Paul Dirac predicted the existence of antimatter in the 1930s, no particle accelerators could reach that energy level, so the experimental evidence of its existence came from cosmic rays.

As the cosmic ray particles move through space, they interact with the cosmic microwave background radiation (CMBR). This microwave energy that permeates the universe is pretty weak, but for the cosmic ray particles, moving at nearly the speed of light, the CMBR appears to be highly energetic. (This is an effect of relativity, because energy is related to motion.)

In 1966, Soviet physicists Georgiy Zatsepin and Vadim Kuzmin, as well as the independent work of Kenneth Greisen of Cornell University, revealed that these collisions would have enough energy to create particles called *mesons* (specifically called pi-mesons, or *pions*). The energy used to create the pions had to come from somewhere (because of conservation of energy), so the cosmic rays would lose energy. This placed an upper bound on how fast the cosmic rays could, in principle, travel.

In fact, the *GZK cutoff energy* needed to create the pions would be about 10^{19} eV (about one-billionth of the Planck energy of 10^{19} GeV).

The problem is that, while most cosmic ray particles fall well below this threshold, some very rare events that have had *more* energy than this threshold — around 10^{20} eV. The most famous of these observations was in 1991 at the University of Utah's Fly's Eye cosmic ray observatory on the U.S. Army's Dugway Proving Ground.

Research since then indicates that the GZK cutoff does indeed exist. The rare occurrence of particles above the cutoff is a reflection of the fact that, very occasionally, these particles reach Earth before they come in contact with enough CMBR photons to slow them down to the cutoff point.

These observations are in conflict with Japan's Akeno Giant Air Shower Array (AGASA) project, which identified nearly ten times as many of these events. The AGASA results implied a potential failure of the cutoff, which could have had implications for a breakdown in relativity, but these other findings decrease the probability of this explanation.

Still, the occasional existence of such energetic particles provides one means of exploring these energy ranges, well above what current particle accelerators could reach, so string theory may have a chance of an experimental test using high-energy cosmic rays, even if they are incredibly rare.

Analyzing dark matter and dark energy

One other astronomical possibility to get results to support string theory comes from the two major mysteries of the universe: dark matter and dark energy. These concepts are discussed at length in Chapters 9 and 14.

The most obvious way that dark matter could help string theory is if it's found that the dark matter is actually supersymmetric particles, such as the photino (the superpartner of the photon) and other possible particles.

Another dark matter possibility is a theoretical particle called an *axion,* originally developed outside of string theory as a means of conserving certain symmetry relationships in quantum chromodynamics. Many string theories contain the axion, so it could be a possibility as well, although the properties suggested don't really match what cosmologists are looking for.

Some of the most significant work in cosmology and astrophysics today are attempts to detect dark matter, and there seems to be a lot of it in the universe. So there's some hope that physicists will make headway on its composition within the foreseeable future.

Detecting cosmic superstrings

Cosmic strings (which in this case are not the same things as the fundamental superstrings of string theory) were originally proposed in 1976 by Tom Kibble of Imperial College London, who suggested that in the aftermath of the big bang, as the universe went through a rapid cooling phase, defects may have remained behind. These defects in quantum fields are similar to when you rapidly freeze water into ice, creating a white substance that is full of defects.

For a while in the 1980s, some scientists thought cosmic strings might be the original seed material for galaxies, but the CMBR data doesn't indicate this to be true. Years later, string theory would resurrect the notion of cosmic strings in a new form.

According to some string theory models, superstrings created in the big bang may have expanded along with the universe itself, creating *cosmic superstrings.* An alternate explanation explains these cosmic superstrings as remnants from the collision of two branes.

Cosmic superstrings would be incredibly dense objects. Narrower than a proton, a single meter of a cosmic superstring could weigh about the same as North America. As they vibrated in space, they could generate massive gravity waves rippling out through space-time.

One way of seeing the cosmic superstrings would be through the *gravitational lensing,* where the string's gravity bends the light of a star, as shown in Figure 12-2. This might mean that we see one star in two different locations, each equally bright.

Figure 12-2:
Gravity from a cosmic superstring could bend the light from a star.

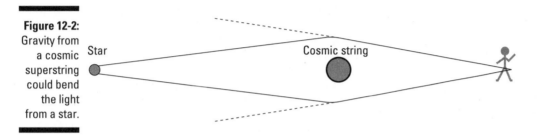

According to Joe Polchinski, the best way to look for cosmic superstrings is to observe pulsars (such as the ones that Fermi is detecting, as mentioned earlier in this chapter). Pulsars are like astronomical lighthouses, spinning as they fire regular beams of electromagnetic radiation into the universe, which follow a predictable pattern. The gravity from a cosmic superstring could cause ripples in space-time that alter this pattern in a way that should be detectable here on Earth.

Looking for Evidence Closer to Home: Using Particle Accelerators

Although it would be nice if nature gave us the experimental results we need, scientists are never content to wait for a lucky break, which is why they proceed with experiments in apparatuses that they control. For high-energy particle physics, this means particle accelerators.

A *particle accelerator* is a device that uses powerful magnetic fields to accelerate a beam of charged particles up to incredibly fast speeds and then collides it with a beam of particles going the other way. Scientists can then analyze the results of the collision.

Relativistic Heavy Ion Collider (RHIC)

The Relativistic Heavy Ion Collider (RHIC) is a particle accelerator at Brookhaven National Laboratory in New York. It went online in 2000, after a decade of planning and construction.

The RHIC name comes from the fact that it accelerates heavy ions — that is, atomic nuclei stripped of their electrons — at relativistic speeds (99.995 percent the speed of light) and then collides them. Because the particles are atomic nuclei, the collisions contain a lot of power in comparison to pure proton beams (though it also takes more time and energy to get them up to that speed).

By slamming two gold nuclei together, physicists can obtain a temperature 300 million times hotter than the sun's surface. The protons and neutrons that normally make up the nuclei of gold break down at this temperature into a plasma of quark and gluons.

This *quark-gluon plasma* is predicted by quantum chromodynamics (QCD), but the problem is that the plasma is supposed to behave like a gas. Instead, it behaves like a liquid. According to Leonard Susskind, string theory may be able to explain this behavior using a variation on the Maldacena conjecture (described in Chapter 11). In this way, the quark-gluon plasma may be described by an equivalent theory in the higher-dimensional universe: a black hole, in this case!

These results are far from conclusive, but theorists are looking at the behavior of these collisions to find ways to apply string theory to make greater sense of the existing physical models (QCD in this case), which is a powerful tool to help gain support of string theory.

Large Hadron Collider (LHC)

The Large Hadron Collider (LHC) is a massive apparatus, built underground at the CERN particle physics facility on the border of Switzerland and France. (CERN is the European particle physics center that was, in 1968, the birthplace of string theory.) The accelerator itself is about 27 kilometers (17 miles) in circumference, as shown in Figure 12-3. The 9,300 magnets of the facility can accelerate protons into collisions up to possibly 14 trillion electron volts (TeV), well beyond our current experimental limitations. The cost of the LHC was around $9 billion as of this writing.

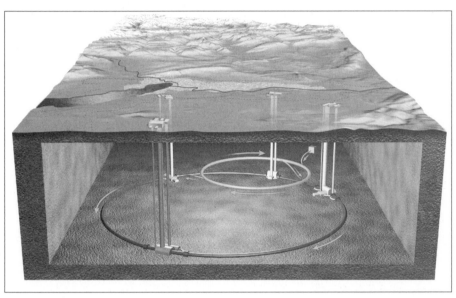

Courtesy CERN Press Office

Figure 12-3:
The Large
Hadron
Collider
is built in
a circular
tunnel that
is 17 miles
around.

On September 10, 2008, the LHC came online by officially running a beam
the full length of the tunnel. On September 19, a faulty electrical connection
caused a rupture in the vacuum seal, resulting in a leak of 6 tons of liquid
helium. The repairs (and upgrades to avoid the problem in the future) were
scheduled to take at least a year.

Due to this, there have yet to be any significant experimental results from the
LHC, but they should be coming in the next year or so. The 14 TeV energy
level might be able to reach several possible experimental results:

- ✓ Microscopic black holes, which would support predictions of extra
  dimensions

- ✓ Supersymmetric particle (sparticle) creation

- ✓ Experimental confirmation of the Higgs boson, the final Standard Model
  particle to remain unobserved

- ✓ Evidence of curled-up extra dimensions

One of the greatest pieces of evidence for string theory could actually be
a lack of evidence. If the experiments at the LHC register some "missing
energy," a couple of possibilities could provide amazing support for string
theory.

 ✔ First, the collisions could create new sparticles that form dark matter, which then flows out of the facility without interacting with the normal matter (like, you know, the detector itself).

 ✔ Second, missing energy could result from energy (or sparticles) that are actually traveling directly into the extra dimensions, rather than into our own 4-dimensional space-time.

Either of these findings would be a great discovery, and either the supersymmetric particles or extra dimensions would have profound implications for string theory.

Colliders of the future

Particle accelerators are so massive that there are no set designs for them; each particle accelerator is its own prototype. The next one on the books appears to be the International Linear Collider (ILC), which is an electron-positron collider. One benefit of this is that electrons and positrons, because they're fundamental particles and not composite particles like protons, are a lot less messy when they collide.

The ILC has not been approved. Proposals, including location, could be voted on around 2012 and, if approved, it could be running in the late 2010s. Early estimates for the project give a minimum cost of $6.65 billion (excluding little things like actually buying the land and other incidental costs).

It's also possible that the LHC might be the last of the large particle accelerators, because new proposed technologies may be developed that allow for rapid particle acceleration that doesn't require massive facilities.

One such design, proposed at CERN, is the Compact Linear Collider (CLIC). The CLIC would use a new two-beam accelerator, where one beam accelerates a second beam. The energy from a low-energy (but high-current) beam into a high-energy (but low-current) beam could allow for accelerations up to 5 TeV in a much shorter distance than traditional accelerators. A decision on CLIC could be made in 2010, with construction probably completed shortly after 2020.

Part IV
The Unseen Cosmos: String Theory On the Boundaries of Knowledge

The 5th Wave   By Rich Tennant

"Along with 'Antimatter,' and 'Dark Matter,' we've recently discovered the existence of 'Doesn't Matter,' which appears to have no effect on the universe whatsoever."

In this part . . .

String theory brings up many amazing possibilities about how to explain the fundamental properties of our universe, such as space, time, and matter itself. Though physicists are far from reaching a final version of string theory, there are many possible implications worth thinking about, even at this early stage in the theory's development.

In this part, I explore the implications of string theory on our view of the universe. I explain how mathematicians and scientists use the concept of dimensions and how the extra dimensions in string theory can be interpreted. Then I return to the ideas of cosmology and show how string theory presents possible explanations for properties in our universe.

String theory can also be used as a means of presenting the ideas of other universes, some of which may in theory someday be accessible. Finally, I discuss the possibility of whether string theory could ever allow for time travel.

Chapter 13

Making Space for Extra Dimensions

*O*ne of the most fascinating aspects of string theory is the requirement of extra dimensions to make the theory work. String theory requires nine space dimensions, while M-theory seems to require ten space dimensions. Under some theories, some of these extra dimensions may actually be long enough to interact with our own universe in a way that could be observed.

In this chapter, you get a chance to explore and understand the meaning of these extra dimensions. First, I introduce the concept of dimensions in a very general way, talking about different approaches mathematicians have used to study 2- and 3-dimensional space. Then I tackle the idea of time as the fourth dimension. I analyze the ways in which the extra dimensions may manifest in string theory and whether the extra dimensions are really necessary.

What Are Dimensions?

Any point in a mathematical space can be defined by a set of coordinates, and the number of coordinates required to define that point is the number of dimensions the space possesses. In a 3-dimensional space like you're used to, for example, every point can be uniquely defined by precisely three coordinates — three pieces of information (length, width, and height). Each dimension represents a degree of freedom within the space.

Though I've been talking about dimensions in terms of space (and time), the concept of dimensions extends far beyond that. For example, the matchmaking Web site eHarmony.com provides a personality profile that claims to assess you on 29 dimensions of personality. In other words, it uses 29 pieces of information as parameters for its dating matches.

I don't know the details of eHarmony's system, but I have some experience with using dimensions on other dating sites. Say you wanted to find a potential romantic partner. You're trying to target a specific type of person by entering different pieces of information: gender, age range, location, annual income, education level, number of kids, and so on. Each of these pieces of information narrows down the "space" that you're searching on the dating site. If you have a complete space consisting of every single person who has a profile on the dating site, when your search is over you're narrowed to searching only among those who are within the ranges that you've specified.

Say Jennifer is a female, age 30, in Dallas, with a college degree and one child. Those coordinates "define" Jennifer (at least to the dating site), and searches that sample those coordinates will include Jennifer as one of the "points" (if you think of each person as a point) in that section of the space.

The problem with this analogy is that you end up with a large number of points within the dating site space that have the same coordinates. There may be another girl, Andrea, who enters essentially identical information as Jennifer. Any search of the sample space that brings up Jennifer also brings up Andrea. In the physical space that we live in, each point is unique.

Each dimension — in both mathematics and in the dating site example — represents a *degree of freedom* within the space. By changing one of the coordinates, you move through the space along one of the dimensions. For example, you can exercise a degree of freedom to search for someone with a different educational background or a different age range or both.

When scientists talk about the number of dimensions in string theory, they mean the degrees of freedom required for these theories to work without going haywire. In Chapter 10, I explain that the bosonic string theory required 25 space dimensions to be consistent. Later, superstring theory required 9 space dimensions. M-theory seems to require 10 space dimensions, and the later F-theory includes 12 total dimensions.

2-Dimensional Space: Exploring the Geometry of Flatland

Many people think of geometry (the study of objects in space) as a flat, 2-dimensional space that contains two degrees of freedom — up or down and right or left. Throughout most of modern history, this interest has been the study of Euclidean geometry or Cartesian geometry.

Euclidean geometry: Think back to high school geometry

Probably the most famous mathematician of the ancient world was Euclid, who has been called the father of geometry. Euclid's 13-volume book, *Elements,* is the earliest known book to have taken all the existing knowledge of geometry at the time of its writing (around 300 BCE). For nearly 2,000 years, virtually all of geometry could be understood just by reading *Elements,* which is one reason why it was the most successful math book ever.

In *Elements,* Euclid started off presenting the principles of plane geometry — that is, the geometry of shapes on a flat surface, as in Figure 13-1. An important consequence of Euclidean plane geometry is that if you take the measure of all three angles inside of a triangle, they add up to 180 degrees.

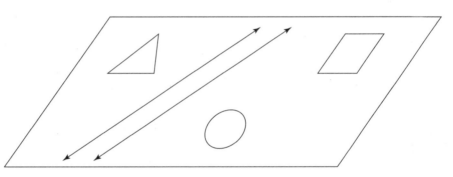

Figure 13-1: In Euclidean geometry, all figures are flat, as if drawn on a sheet of paper.

Later in the volumes, Euclid extended into 3-dimensional geometry of solid objects, such as cubes, cylinders, and cones. The geometry of Euclid is the geometry typically taught in school to this day.

Cartesian geometry: Merging algebra and Euclidean geometry

Modern analytic geometry was founded by French mathematician and philosopher Rene Descartes, when he placed algebraic figures on a physical grid. This sort of Cartesian grid is shown in Figure 13-2. By applying concepts from Euclidean geometry to the equations depicted on the grids, insights into geometry and algebra could be obtained.

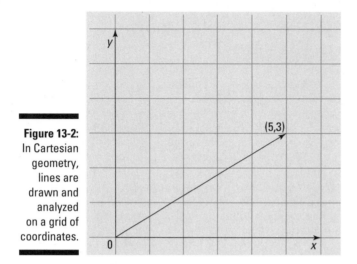

Figure 13-2:
In Cartesian
geometry,
lines are
drawn and
analyzed
on a grid of
coordinates.

Around the same time that Galileo was revolutionizing the heavens,
Descartes was revolutionizing mathematics. Until his work, the fields of algebra and geometry were separate. His idea was to display algebraic equations
graphically, providing a way to translate between geometry and algebra.

Books of many dimensions

The book *Flatland: A Romance of Many
Dimensions* by Edwin A. Abbott, written in 1884,
is a classic in the mathematics community for
explaining the concept of multiple dimensions.
In this book, A. Square lives in a flat world
and gains perspective when he encounters a
sphere passing through his world who pulls
him out of it so he can briefly experience three
dimensions.

Flatland appears to have been part of a growing popular culture interest in extra dimensions
during the late 1800s. Lewis Carroll had written
a story in 1865 entitled "Dynamics of a Particle,"
which included 1-dimensional beings on a flat
surface, and the idea of space going crazy is
clearly a theme in Carroll's *Alice's Adventures*

in Wonderland (1865) and *Through the Looking
Glass* (1872). Later, H. G. Wells used the concepts of extra dimensions in several stories,
most notably in *The Time Machine* (1895),
where time is explicitly described as the fourth
dimension a full decade before Einstein presented the first inkling of relativity.

Various independent sequels have been written to *Flatland* through the years to expand on
the concept. These include Dionys Burger's
Sphereland (1965), Ian Stewart's *Flatterland*
(2001), and Rudy Rucker's *Spaceland* (2002). A
related book is the 1984 science-fiction novel
The Planiverse, where scientists in our world
establish communication with a Flatland-like
world.

Using the Cartesian grid, you can define a line by an equation; the line is the set of solutions to the equation. In Figure 13-2, the line goes from the origin to the point (5, 3). Both the origin (0, 0) and (5, 3) are correct solutions to the equation depicted by the line (along with all the other points on the line).

Because the grid is 2-dimensional, the space that the grid represents contains two degrees of freedom. In algebra, the degrees of freedom are represented by variables, meaning that an equation that can be shown on a 2-dimensional surface has two variable quantities, often *x* and *y*.

Three Dimensions of Space

When looking in our world, it has three dimensions — up and down, left and right, back and forth. If you give a longitude, latitude, and an altitude, you can determine any location on Earth, for example.

A straight line in space: Vectors

Expanding on the idea of Cartesian geometry, you find that it's possible to create a Cartesian grid in three dimensions as well as two, as shown in Figure 13-3. In such a grid, you can define an object called a *vector,* which has both a direction and a length. In 3-dimensional space, every vector is defined by three quantities.

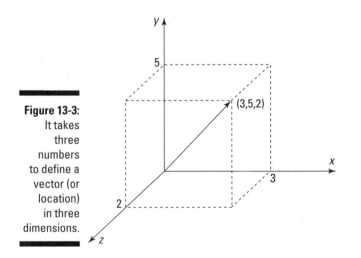

Figure 13-3:
It takes three numbers to define a vector (or location) in three dimensions.

Vectors can, of course, exist in one, two, or more than three dimensions. (Technically, you can even have a zero-dimensional vector, although it will always have zero length and no direction. Mathematicians call such a case "trivial.")

Treating space as containing a series of straight lines is probably one of the most basic operations that can take place within a space. One early field of mathematics that focuses on the study of vectors is called *linear algebra,* which allows you to analyze vectors and things called *vector spaces* of any dimensionality. (More advanced mathematics can cover vectors in more detail and extend into nonlinear situations.)

One of the major steps of working with vector spaces is to find the *basis* for the vector space, a way of defining how many vectors you need to define *any* point in the entire vector space. For example, a 5-dimensional space has a basis of five vectors. One way to look at superstring theory is to realize that the directions a string can move can only be described with a basis of ten distinct vectors, so the theory describes a 10-dimensional vector space.

Twisting 2-dimensional space in three dimensions: The Mobius strip

In the classic book *Flatland,* the main character is a square (literally — he has four sides of equal length) who gains the ability to experience three dimensions. Having access to three dimensions, you can perform actions on a 2-dimensional surface in ways that seem very counterintuitive. A 2-dimensional surface can actually be twisted in such a way that it has no beginning and no end!

The best known case of this is the *Mobius strip,* shown in Figure 13-4. The Mobius strip was created in 1858 by German mathematicians August Ferdinand Mobius and Johann Benedict Listing.

Figure 13-4:
A Mobius strip is twisted so it has only one continuous surface.

You can create your own Mobius strip by taking a strip of paper — kind of like a long bookmark — and giving it a half-twist. Then take the two ends of the strip of paper and tape them together. Place a pencil in the middle of the surface and draw a line along the length of the strip without taking your pencil off the paper.

A curious thing happens as you continue along. Eventually, without taking your pencil from the paper, the line is drawn on every part of the surface and eventually meets up with itself. There is no "back" of the Mobius strip, which somehow avoids the pencil line. You've drawn a line along the entire shape without lifting your pencil.

In mathematical terms (and real ones, given the result of the pencil experiment), the Mobius strip has only one surface. There is no "inside" and "outside" of the Mobius strip, the way there is on a bracelet. Even though the two shapes may look alike, they are mathematically very different entities.

The Mobius strip does, of course, have an end (or boundary) in terms of its width. In 1882, the German mathematician Felix Klein expanded on the Mobius strip idea to create a *Klein bottle:* a shape that has no inside or outside surface, but also has no boundaries in any direction. Take a look at Figure 13-5 to understand the Klein bottle. If you traveled along the "front" of the path (with the x's), you'd eventually reach the "back" of that path (with the o's).

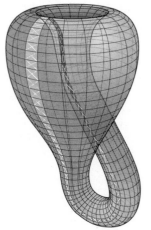

Figure 13-5:
A Klein bottle has no boundary (edge).

If you were an ant living on a Mobius strip, you could walk its length and eventually get back to where you started. Walking its width, you'd eventually run into the "edge of the world." An ant living on a Klein bottle, however, could go in any direction and, if it walked long enough, eventually find itself back where it started. (Traveling along the o path eventually leads back to the x's.) The difference between walking on a Klein bottle and walking on a sphere is that the ant wouldn't just walk along the outside of the Klein bottle, like it would on a sphere, but it would cover both surfaces, just like on the Mobius strip.

More twists in three dimensions: Non-Euclidean geometry

The fascination with strange warping of space in the 1800s was perhaps nowhere as clear as in the creation of *non-Euclidean geometry,* where mathematicians began to explore new types of geometry that weren't based on the rules laid out 2,000 years earlier by Euclid. One version of non-Euclidean geometry is Riemannian geometry, but there are others, such as projective geometry.

The reason for the creation of non-Euclidean geometry is based in Euclid's *Elements* itself, in his "fifth postulate," which was much more complex than the first four postulates. The fifth postulate is sometimes called the *parallel postulate* and, though it's worded fairly technically, one consequence is important for string theory's purposes: A pair of parallel lines never intersects.

Well, that's all well and good on a flat surface, but on a sphere, for example, two parallel lines can and do intersect. Lines of longitude — which are parallel to each other under Euclid's definition — intersect at both the north and south poles. Lines of latitude, also parallel, don't intersect at all. Mathematicians weren't sure what a "straight line" on a circle even meant!

One of the greatest mathematicians of the 1800s was Carl Friedrich Gauss, who turned his attention to ideas about non-Euclidean geometry. (Some earlier thoughts on the matter had been kicked around over the years, such as those by Nikolai Lobachevsky and Janos Bolyai.) Gauss passed the majority of the work off to his former student, Bernhard Riemann. Riemann worked out how to perform geometry on a curved surface — a field of mathematics called *Riemannian geometry.* One consequence — that the angles of a triangle do *not* add up to 180 degrees — is depicted in Figure 13-6.

Figure 13-6: Sometimes the angles of a triangle don't measure up to 180 degrees.

The mathematics of artwork

Understanding and manipulating space is a key feature of artwork, which often attempts to reflect a 3-dimensional reality on a 2-dimensional surface. This is probably most notable in the work of Pablo Picasso and M. C. Escher, where space has been manipulated in such a way that the manipulation itself is part of the artistic message.

Most artists try to manipulate space so it's not noticed. One of the most common examples of this is perspective, developed during the Renaissance, which involves creating an image that matches the way the eye perceives space and distance. Parallel railroad tracks appear to meet at the horizon, though they never meet in reality. On a 2-dimensional surface, the basis for the railroad tracks is a triangle that does, in fact, have a corner at the horizon line.

This is precisely the basis of the mathematical field of non-Euclidean geometry called *projective geometry,* where you take one 2-dimensional space and project it in a precise mathematical way onto a second surface. There is an exact 1-to-1 correspondence between the two spaces, even though they look completely different. The two images represent different mathematical ways of looking at the same physical space — one of them an infinite space and one a finite space.

When Albert Einstein developed general relativity as a theory about the geometry of space-time, it turned out that Riemannian geometry was exactly what he needed.

Four Dimensions of Space-Time

In Einstein's general theory of relativity, the three space dimensions connect to a fourth dimension: time. The total package of four dimensions is called *space-time,* and in this framework, gravity is seen as a manifestation of space-time geometry. The story of relativity is told in Chapter 6, but some dimension-related points are worth revisiting.

Hermann Minkowski, not Albert Einstein, realized that relativity could be expressed in a 4-dimensional space-time framework. Minkowski was one of Einstein's old teachers, who had called him a "lazy dog," but he clearly saw the brilliance of relativity.

In a 1908 talk entitled "Space and Time," Minkowski first broached the topic of creating a dimensional framework of space-time (also sometimes called a "Minkowski space"). The Minkowski diagrams, introduced in Chapter 6, are an attempt to graphically represent this 4-dimensional space on a 2-dimensional Cartesian grid. Each point on the grid is a "space-time event," and understanding the ways these events relate to each other is the goal of analyzing relativity in this way.

Even though time is a dimension, it's fundamentally different from the space dimensions. Mathematically, you can generally exchange "left" for "up" and end up with results that are fairly consistent. If you, however, exchange "left one meter" for "one hour from now," it doesn't work out so well. Minkowski divided the dimensions into *spacelike dimensions* and *timelike dimensions*. One spacelike dimension can be exchanged for another, but can't be exchanged with a timelike dimension. (In Chapter 16, you find out about some ideas regarding extra timelike dimensions in our universe.)

The reason for this distinction is that Einstein's equations are written in such a way that they result in a term defined by the space dimensions squared minus a term defined by the time dimension squared. (Because the terms are squared, each term has to be positive, no matter what the value of the dimension.) The space dimensional values can be exchanged without any mathematical problem, but the minus sign means that the time dimension can't be exchanged with the space dimensions.

Adding More Dimensions to Make a Theory Work

For most interpretations, superstring theory requires a large number of extra space dimensions to be mathematically consistent: M-theory requires ten space dimensions. With the introduction of branes as multidimensional objects in string theory, it becomes possible to construct and imagine wildly creative geometries for space — geometries that correspond to different possible particles and forces. It's unclear, at present, whether those extra dimensions exist in a real sense or are just mathematical artifacts.

The reason string theory requires extra dimensions is that trying to eliminate them results in much more complicated mathematical equations. It's not impossible (as you see later in this chapter), but most physicists haven't pursued these concepts in a great deal of depth, leaving science (perhaps by default) with a theory that requires many extra dimensions.

As I mention earlier, from the time of Descartes, mathematicians have been able to translate between geometric and physical representations. Mathematicians can tackle their equations in virtually any number of dimensions that they choose, even if they can't visually picture what they're talking about.

One of the tools mathematicians use in exploring higher dimensions is analogy. If you start with a zero-dimensional point and extend it through space, you get a 1-dimensional line. If you take that line and extend it into a second dimension, you end up with a square. If you extend a square through a third dimension, you end up with a cube. If you then were to take a cube and extend into a fourth dimension, you'd get a shape called a *hypercube*.

A line has two "corners" but extending it to a square gives four corners, while a cube has eight corners. By continuing to extend this algebraic relationship, a hypercube would be a 4-dimensional object with 16 corners, and a similar relationship can be used to create analogous objects in additional dimensions. Such objects are obviously well outside of what our minds can picture.

Humans aren't psychologically wired to be able to picture more than three space dimensions. A handful of mathematicians (and possibly some physicists) have devoted their lives to the study of extra dimensions so fully that they may be able to actually picture a 4-dimensional object, such as a hypercube. Most mathematicians can't (so don't feel bad if you can't).

Whole fields of mathematics — linear algebra, abstract algebra, topology, knot theory, complex analysis, and others — exist with the sole purpose of trying to take abstract concepts, frequently with large numbers of possible variables, degrees of freedom, or dimensions, and make sense of them.

These sorts of mathematical tools are at the heart of string theory. Regardless of the ultimate success or failure of string theory as a physical model of reality, it has motivated mathematics to grow and explore new questions in new ways, and for that alone, it has proved useful.

Sending Space and Time on a Bender

Space-time is viewed as a smooth "fabric," but that smooth fabric can be bent and manipulated in various ways. In relativity, gravity bends our four space-time dimensions, but in string theory more dimensions are bound up in other ways. In relativity and modern cosmology, the universe has an inherent curvature.

The typical approach to string theory's extra dimensions has been to wind them up in a tiny, Planck length–sized shape. This process is called *compactification*. In the 1980s, it was shown that the extra six space dimensions of superstring theory could be compactified into Calabi-Yau spaces.

Since then, other methods of compactification have been offered, most notably G2 compactification, spin-bundle compactification, and flux compactification. For the purposes of this book, the details of the compactification don't matter.

The wraparound universe

Some cosmologists have considered some extreme cases of space warping in our own universe, theorizing that the universe may be smaller than we think. A new field of cosmology called *cosmic topology* attempts to use mathematical tools to study the overall shape of the universe.

In his 2008 book, *The Wraparound Universe*, cosmologist Jean-Pierre Luminet proposes the idea that our universe wraps around so it has no particular boundary, sort of like the Klein bottle in Figure 13-5. Any direction you look, you may be seeing an illusion, as if you were standing in a funhouse full of mirrors that appeared to go on forever. Distant stars may actually be closer than expected, but the light travels a larger path along the wraparound universe to reach us.

In this sort of a scenario, the horizon problem from Chapter 9 ceases to be an issue because the universe is small enough to have become uniform within the timeframe of our universe's existence. Inflation is consistent with the wraparound universe hypothesis, but many of the problems it fixes are solved in other ways.

To picture compactification, think of a garden hose. If you were an ant living on the hose, you'd live on an enormous (but finite) universe. You can walk very far in either of the length directions, but if you go around the curved dimension, you can only go so far. However, to someone very far away, your dimension — which is perfectly expansive at your scale — seems like a very narrow line with no space to move except along the length.

This is the principle of compactification — we can't see the extra universes because they're so small that nothing we can do can ever distinguish them as a complex structure. If we got close enough to the garden hose, we'd realize that something was there, but scientists can't get close to the Planck length to explore extra compactified dimensions.

Of course, some recent theories have proposed that the extra dimensions may be larger than the Planck length and theoretically in the range of experiment.

Still other theories exist in which our region of the universe only manifests four dimensions, even though the universe as a whole contains more. Other regions of the universe may exhibit additional dimensions. Some radical theories even suppose that the universe as a whole is curved in strange ways.

Are Extra Dimensions Really Necessary?

Though string theory implies extra dimensions, that doesn't mean that the extra dimensions need to exist as dimensions of space. Some work has been done to formulate a 4-dimensional string theory where the extra degrees of freedom aren't physical space dimensions; but the results are incredibly complex, and it doesn't seem to have caught on.

Several groups have performed this sort of work, because some physicists are uncomfortable with the extra space dimensions that seem to be required by string theory. In the late 1980s, a group worked on an approach called free fermions. Other approaches that avoid introducing additional dimensions include the covariant lattice technique, asymmetric orbifolds, the 4-D *N*=2 string (what's in a name?), and non-geometric compactifications. These are technically complex formulations of string theory (aren't they all?) that seem to be ignored by virtually all popular books on the subject, which focus on the idea of extra dimensions to the exclusion of these alternative approaches. Even among string theorists, the geometric approach of compactifying extra dimensions is the dominant approach.

One early, technically complex (and largely ignored) approach to 4-dimensional string theory is work performed by S. James Gates Jr., of the University of Maryland at College Park (along with assistance from Warren Siegel of Stony Brook University's C. N. Yang Institute for Theoretical Physics). This work is by no means the dominant approach to 4-dimensional string theory, but it's benefit is that it can be explained and understood (in highly simplified terms) without a doctorate in theoretical physics.

Offering an alternative to multiple dimensions

In Gates's approach, he essentially trades dimensions for charges. This creates a sort of dual approach that's mathematically similar to the approach in extra space dimensions, but doesn't actually require the extra space dimensions nor require guessing at compactification techniques to eliminate the extra dimensions.

This idea dates back to a 1938 proposal by British physicist Nicolas Kemmer. Kemmer proposed that the quantum mechanical properties of charge and spin were different manifestations of the same thing. Specifically, he said that the neutron and proton were identical, except that they rotated differently in some extra dimension, which resulted in a charge on the proton and no charge on the neutron. The resulting mathematics, which analyzes the physical properties of these particles, is called an *isotopic charge space* (originally developed by Werner Heisenberg and Wolfgang Pauli, then used by Kemmer). Though this is an "imaginary space" (meaning that the coordinates are unobservable in the usual sense), the resulting mathematics describes properties of protons and neutrons, and is at the foundation of the current Standard Model.

Gates's approach was to take Kemmer's idea in the opposite direction: If you wanted to get rid of extra dimensions, perhaps you could view them as imaginary and get charges. (The word "charge" in this sense doesn't really mean electrical charge, but a new property to be tracked, like "color charge" in QCD.) The result is to take vibrational dimensions of the heterotic string and view them as "left charge" and "right charge."

When Gates applied this concept to the heterotic string, the trading didn't come out even — to give up six space dimensions, he ended up gaining more than 496 right charges!

In fact, together with Siegel, Gates was able to find a version of heterotic string theory that matched these 496 right charges. Furthermore, their solution showed that the left charges would correspond to the family number. (There are three known generations, or families, of leptons as shown in Figure 8-1 in Chapter 8 — the electron, muon, and tau families. The family number indicates which generation the particle belongs to.)

This may explain why there are multiple families of particles in the Standard Model of particle physics. Based on these results, a string theory in four dimensions could require extra particle families! In fact, it would require many more particle families than the three that physicists have seen. These extra families (if they exist) could include particles that could make up the unseen dark matter in our universe.

Weighing fewer dimensions against simpler equations

The usefulness of these 4-dimensional results is hindered by the sheer complexity of the resulting equations (even by string theory standards). Although all string theories are complex, 4-dimensional string theories have, to date, shown meager predictive power. Assuming the extra dimensions lead to equations that are easier to handle, most physicists choose to work under the assumption of greater numbers of dimensions.

This goes back to the idea that the principle of Occam's razor, which says that a scientist shouldn't make a theory unnecessarily complex. The simplest explanation that fits the facts is the one that physicists tend to gravitate toward.

In this case, Occam's razor cuts both ways. The simpler mathematical equation of 10-dimensional string theory requires stipulating a large number of space dimensions that no one has ever observed, which would certainly seem to go against Occam's razor. But the type of isotopic charge coordinate used in Gates's approach is exactly the same as the ones that provide the mathematical foundations of the Standard Model — where the isotopic dimensions aren't observed.

In the end, the 4-dimensional interpretations of string theory are a powerful way of understanding how complex string theory can be. One of the most basic aspects of string theory has been the idea that it requires extra space dimensions, but this work shows that string theory doesn't necessarily require even that. If these approaches are right, and the degrees of freedom inherent in the theory don't require extra space dimensions, then the physical principles at the heart of string theory may be completely unexpected.

Chapter 14

Our Universe — String Theory, Cosmology, and Astrophysics

. . . .

In This Chapter

▶ Looking back beyond the big bang theory

▶ Tying black holes to string theory

▶ Knowing where the universe has been and where it may be going

▶ Tackling the question of how the universe supports life

. . . .

*T*hough string theory started as a theory of particle physics, much of the significant theoretical work today is in applying the startling predictions of string theory and M-theory to the field of cosmology. Chapter 9 covered some of the amazing facts science has discovered about our universe, especially in the last century.

In this chapter, I return to these same ideas from the background of string theory. I explain how string theory relates to our understanding of the big bang, the theory of the universe's origin. I then discuss what string theory has to say about another mystery of the universe — black holes. From there, I cover what string theory reveals about how the universe changes over time and how it may change in the future. Finally, I return to the question of why the universe seems perfectly tuned to allow for life and what, if anything, string theory (along with the anthropic principle) may have to say about it.

The Start of the Universe with String Theory

According to the big bang theory, if you extrapolate the expanding universe backward in time, the entire known universe would have been compacted down into a singular point of incredibly immense density. It reveals nothing, however, about whether anything existed a moment prior to that point. In fact, under the big bang theory — formulated in a universe of quantum physics and relativity — the laws of physics result in meaningless infinities at that moment. String theory may offer some answers to what came before and what caused the big bang.

What was before the bang?

String theory offers the possibility that we are "stuck" on a brane with three space dimensions. These brane world scenarios, such as the Randall-Sundrum models, offer the possibility that before the big bang something was already here: collections of strings and branes.

The search for an eternal universe

Scientists were originally very upset by the big bang theory, because they believed in an eternal universe, meaning that the universe had no starting point (and, on average, didn't change over time). Einstein believed this, though he abandoned it when evidence suggested otherwise. Fred Hoyle devoted most of his career to trying to prove the universe was eternal. Today, some physicists continue to look for ways to explain what, if anything, existed before the big bang.

Some cosmologists say that the question of what happened at or before the big bang is inherently unscientific, because science currently has no way of extending its physical theories past the singularity at the dawn of our universe's timeline. Others point out that if we never ask the questions, we'll never discover a way to answer them.

Though string theory isn't yet ready to answer such questions, that hasn't stopped cosmologists from beginning to ask the questions and offer possible scenarios. In these scenarios, which are admittedly vague, the pre–big bang universe (which likely is not confined to only three space dimensions) is a conglomerate of p-branes, strings, anti-strings, and anti-p-branes. In many cases, these objects are still "out there" somewhere beyond our own 3-brane, perhaps even impacting our own universe (as in the case of the Randall-Sundrum models).

One of these models was a pre–big bang model presented by Gabriele Veneziano — the same physicist who came up with the 1968 dual resonance model that sparked string theory. In this model, our universe is a black hole in a more massive universe of strings and empty space. Prior to the current expansion phase, there was a period of contraction. Though probably not completely true according to today's major models, this work by Veneziano (and similar ideas by others) has an impact on most of the superstring cosmology work today, because it pictures our known universe as just a subset of the universe, with a vast "out there" beyond our knowledge.

The old-fashioned cyclic universe model

One idea that was popular in the 1930s was that of a *cyclic universe,* in which the matter density was high enough for gravity to overcome the expansion of the universe. The benefit of this model was that it allowed the big bang to be correct, but the universe could still be eternal.

In this cyclic model, the universe would expand until gravity began to pull it back, resulting in a "big crunch" where all matter returned to the primordial "superatom" — and then the cycle of expansion would start all over again.

The problem is that the second law of thermodynamics dictates that the entropy, or disorder, in the universe would grow with each cycle. If the universe went through an infinite number of cycles, the amount of disorder in the universe would be infinite — every bit of the universe would be in thermal equilibrium with every other bit of the universe. In a universe where every region has exactly the same structure, no one region has more order than any other, so all regions have the maximum amount of disorder allowed. (If the universe had gone through a finite number of cycles, scientists still ran into the problem of how the whole thing started; they just pushed it back a few cycles. This kind of defeated the whole purpose of the model, so the model assumed an infinite number of cycles.)

String theory, however, might just have a way of bringing back the cyclic model in a new form.

What banged?

The big bang theory doesn't offer any explanation for what started the original expansion of the universe. This is a major theoretical question for cosmologists, and many are applying the concepts of string theory in attempts to answer it. One controversial conjecture is a cyclic universe model called the *ekpyrotic universe theory,* which suggests that our own universe is the result of branes colliding with each other.

The banging of strings

Well before the introduction of M-theory or brane world scenarios, there was a string theory conjecture of why the universe had the number of dimensions we see: A compact space of nine symmetrical space dimensions began expanding in three of those dimensions. Under this analysis, a universe with three space dimensions (like ours) is the most likely space-time geometry.

In this idea, initially posed in the 1980s by Robert Brandenberger and Cumrun Vafa, the universe began as a tightly wound string with all dimensions symmetrically confined to the Planck length. The strings, in effect, bound the dimensions up to that size.

Brandenberger and Vafa argued that in three or fewer dimensions, it would be likely for the strings to collide with anti-strings. (An *anti-string* is essentially a string that winds in a direction opposite the string.) The collision annihilates the string which, in turn, unleashes the dimensions it was confining. They thus begin expanding, as in the inflationary and big bang theories.

Instead of thinking about strings and anti-strings, picture a room that has a bunch of cables attached to random points on the walls. Imagine that the room wants to expand with the walls and floor and ceiling trying to move away from each other — but they can't because of the cables. Now imagine that the cables can move, and every time they intersect, they can recombine. Picture two taut cables stretching from the floor to the ceiling that intersect to form a tall, skinny X. They can recombine to become two loose cables — one attached to the floor and one attached to the ceiling. If these had been the only two cables stretching from floor to ceiling, then after this interaction, the floor and ceiling are free to move apart from each other.

In the Brandenberger and Vafa scenario, this dimension (up-down), as well as two others, are free to grow large. The final step is that in four or more space dimensions, the moving strings will typically never meet. (Think about how points moving in two space dimensions will probably never meet, and the rationale gets extended to higher dimensions.) So this mechanism only works to free three space dimensions from their cables.

In other words, the very geometry of string theory implies that this scenario would lead to us seeing fewer than four space dimensions — dimensions of four or more are less likely to go through the string/anti-string collisions required to "liberate" them from the tightly bound configuration. The higher dimensions continue to be bound up by the strings at the Planck length and are therefore unseen.

With the inclusion of branes, this picture gets more elaborate and harder to interpret. Research into this approach in recent years hasn't been reassuring. Many problems arise when scientists try to embed this idea more rigorously into the mathematics of string theory. Still, this is one of the few explanations of why there are four dimensions that make any sense, so string theorists haven't completely abandoned it as a possible reason for the big bang.

A brane-fueled, 21st-century cyclic model: The ekpyrotic universe

In the ekpyrotic universe scenario, our universe is created from the collision of branes. The matter and radiation of our universe comes from the kinetic energy created by the collision of these two branes.

The ekpyrotic universe scenario was proposed in a 2001 paper by Paul Steinhardt of Princeton University, Burt Ovrut of the University of Pennsylvania, and Neil Turok, formerly of Cambridge University and currently the director of the Perimeter Institute for Theoretical Physics in Waterloo, Ontario, along with Steinhardt's student, Justin Khoury.

The theory builds on the ideas that some M-theory brane world scenarios show that the extra dimensions of string theory may be extended, perhaps even infinite in size. They are also probably not expanding (or at least string theorists have no reason to think they are) the way that our own three space dimensions are. When you play the video of the universe backward in time, these dimensions don't contract.

Now imagine that within these dimensions you have two infinite 3-branes. Some mechanism (such as gravity) draws the branes together through the infinite extra dimensions, and they collide with each other. Energy is generated, creating the matter for our universe and pushing the two branes apart. Eventually, the energy from the collision dissipates and the branes are drawn back together to collide yet again.

The ekpyrotic model is divided into various *epochs* (periods of time), based upon what influences dominate:

- ✔ The big bang
- ✔ The radiation-dominated epoch
- ✔ The matter-dominated epoch
- ✔ The dark energy–dominated epoch
- ✔ The contraction epoch
- ✔ The big crunch

The story up until the contraction epoch is essentially identical to that made by regular big bang cosmology. The radiation that is spawned by the brane collision (the big bang) means the radiation-dominated epoch is fairly uniform (save for quantum fluctuations), so inflation may be unnecessary. After about 75,000 years, the universe becomes a particle soup during the matter-dominated epoch. Today and for many years, we are in the dark energy–dominated epoch, until the dark energy decays and the universe begins contracting once again.

Because the theory involves two branes colliding, some called this the "big splat" theory or the "brane smash" theory, which is certainly easier to pronounce than *ekpyrotic.* The word "ekpyrotic" comes from the Greek word "ekpyrosis," which was an ancient Greek belief that the world was born out of fire. (Burt Ovrut reportedly thought it sounded like a skin disease.)

Some feel that the ekpyrotic universe model has a lot going for it — it solves the flatness and horizon problems like inflationary theory does, while also providing an explanation for why the universe started in the first place — but the creators are still far from proving it. Stephen Hawking has bet Neil Turok that findings from the European Space Agency's Planck satellite will verify the inflationary model and rule out the ekpyrotic model, but Hawking has been known to have to pay out on these sorts of bets in the past (as you can read about in the "String theory and the black hole information paradox" section later in this chapter).

One benefit is that this model avoids the problem of previous cyclic models, because each universe in the cycle is larger than the one before it. Because the volume of the universe increases, the total entropy of the universe in each cycle can increase without ever reaching a state of maximum entropy.

There is obviously much more detail to the ekpyrotic model than I've included here. If you're interested in this fascinating theory, I highly recommend Paul J. Steinhardt and Neil Turok's popular book *Endless Universe: Beyond the Big Bang*. In addition to the lucid and nontechnical discussion of complex scientific concepts, their descriptions offer a glimpse inside the realm of theoretical cosmology, which is well worth the read.

Explaining Black Holes with String Theory

One major mystery of theoretical physics that requires explanation is the behavior of black holes, especially regarding how black holes evaporate and whether they lose information. I introduce these topics in Chapter 9, but with the concepts of string theory in hand, you may be able to further your understanding of them.

Black holes are defined by general relativity as smooth entities, but at very small scales (such as when they evaporate down to the Planck length in size), quantum effects need to be taken into account. Resolving this inconsistency is the sort of thing that string theory should be good at, if it's true.

String theory and the thermodynamics of a black hole

When Stephen Hawking described the Hawking radiation emitted by a black hole, he had to use his physical and mathematical intuition, because quantum physics and general relativity aren't reconciled. One of the major successes of string theory is in offering a complete description of (some) black holes.

Hawking radiation takes place when radiation is emitted from a black hole, causing it to lose mass. Eventually, the black hole evaporates into nothing (or almost nothing).

Stephen Hawking's incomplete argument

Hawking's paper on the way a black hole radiates heat (also called thermodynamics) begins a line of reasoning that doesn't quite work all the way through to the end. In the middle of the proof there's a disconnect, because no theory of quantum gravity exists that would allow the first half of his reasoning (based on general relativity) to connect with the second half of his reasoning (based on quantum mechanics).

The reason for the disconnect is that performing a detailed thermodynamics analysis of a black hole involves examining all the possible quantum states of the black hole. But black holes are described with general relativity, which treats them as smooth — not quantum — objects. Without a theory of quantum gravity, there seems to be no way to analyze the specific thermodynamic nature of a black hole.

In Hawking's paper, this connection was made by means of his intuition, but not in the sense that most of us probably think of intuition. The intuitive leap he took was in proposing precise mathematical formulas, called *greybody factors,* even though he couldn't absolutely prove where they came from.

Most physicists agree that Hawking's interpretation makes sense, but a theory of quantum gravity would show whether a more precise process could take the place of his intuitive step.

String theory may complete the argument

Work by Andrew Strominger and Cumrun Vafa on the thermodynamics of black holes is seen by many string theorists as the most powerful evidence in support of string theory. By studying a problem that is mathematically equivalent to black holes — a dual problem — they precisely calculated the black hole's thermodynamic properties in a way that matched Hawking's analysis.

Sometimes, instead of simplifying a problem directly, you can create a *dual problem,* which is essentially identical to the one you're trying to solve but is much simpler to handle. Strominger and Vafa used this tactic in 1996 to calculate the entropy in a black hole.

In their case, they found that the dual problem of a black hole described a collection of 1-branes and 5-branes. These "brane constructions" are objects that can be defined in terms of quantum mechanics. They found that the results matched precisely with the result Hawking anticipated 20 years earlier.

Now, before you get too excited, the Strominger and Vafa results only work for certain very specific types of black holes, called *extremal black holes.* These extremal black holes have the maximum amount of electric or magnetic charge that is allowed without making the black hole unstable. An extremal black hole has the odd property of possessing entropy but no heat or temperature. (Entropy is a measure of disorder, often related to heat energy, within a physical system.)

At the same time Strominger and Vafa were performing their calculations, Princeton student Juan Maldacena was tackling the same problem (along with thesis advisor Curt Callan). Within a few weeks of Strominger and Vafa, they had confirmed the results and extended the analysis to black holes that are *almost* extremal. Again, the relationship holds up quite well between these brane constructions and black holes, and analyzing the brane constructions yields the results Hawking anticipated for black holes. Further work has expanded this work to even more generalized cases of black holes.

To get this analysis to work, gravity has to be turned down to zero, which certainly seems strange in the case of a black hole that is, quite literally, defined by gravity. Turning off the gravity is needed to simplify the equations and obtain the relationship. String theorists conjecture that by ramping up the gravity again you'd end up with a black hole, but string theory skeptics point out that without gravity you really don't have a black hole.

Still, even a skeptic can't help but think that there must be some sort of relationship between the brane constructions and the black holes because they both follow the Hawking thermodynamics analysis created 20 years earlier. What's even more amazing is that string theory wasn't designed to solve this specific problem, yet it did. The fact that the result falls out of the analysis is impressive, to say the least.

String theory and the black hole information paradox

One of the important aspects of the thermodynamics of black holes relates to the black hole information paradox. This paradox may well have a solution in string theory, either in the string theory analyses described in the previous section or in the holographic principle.

Hawking had said that if an object falls into a black hole, the only information that is retained are the quantum mechanical properties of mass, spin, and charge. All other information was stripped away.

The problem with this is that quantum mechanics is built on the idea that information can't be lost. If information can be lost, then quantum mechanics isn't a secure theoretical structure. Hawking, as a relativist, was more concerned with maintaining the theoretical structure of general relativity, so he was okay with the information being lost if it had to be.

The reason that this lost information is such a major issue for quantum mechanics once again ties into thermodynamics. In quantum mechanics, information is related to the thermodynamic concept of "order." If information is lost, then order is lost — meaning that entropy (disorder) is increased. This means that the black holes would begin generating heat, rising up to billions of billions of degrees in mere moments. Though Leonard Susskind and others realized this in the mid-1980s, they couldn't find the flaws in Hawking's reasoning that would prove him wrong.

In 2004, after a debate that lasted more than 20 years, Hawking announced that he no longer believed this information was forever lost to the universe. In doing so, he lost a 1997 bet with physicist John Preskill. The payoff was a baseball encyclopedia, from which information could be retrieved easily. And who said physicists didn't have a sense of humor?

One reason for Hawking's change of mind was that he redid some of his earlier calculations and found that it was possible that, as an object fell into a black hole, it would disturb the black hole's radiation field. The information about the object could seep out, though probably in mangled form, through the fluctuations in this field.

Another way to approach the problem of black hole information loss is through the holographic principle of Leonard Susskind and Gerard 't Hooft, or the related AdS/CFT correspondence developed by Juan Maldacena. (Both of these principles are discussed in Chapter 11.) If these principles hold for black holes, it may be possible that all the information within the black hole is also encoded in some form on the surface area of the black hole.

 The controversy over the black hole information paradox is described in detail in Susskind's 2008 book, *The Black Hole War: My Battle with Stephen Hawking to Make the World Safe for Quantum Mechanics.*

Still one other approach is to look at the potential multiverse. It's possible that the information that enters a black hole is, in some way, passed from this universe into a parallel universe. I cover this intriguing possibility in Chapter 15.

The Evolution of the Universe

Other questions that scientists hope string theory can answer involve the way the universe changes over time. The brane world scenarios described earlier in this book offer some possibilities, as do the various concepts of a multiverse. Specifically, string theorists hope to understand the reason for the increased expansion of our universe as defined by dark matter and energy.

The swelling continues: Eternal inflation

Some cosmologists have worked hard on a theory called *eternal inflation,* which helps contribute to the idea of a vast multiverse of possible universes, each with different laws (or different solutions to the same law, to be precise).

In eternal inflation, island universes spring up and disappear throughout the universe, spawned by the very quantum fluctuations of the vacuum energy itself. This is seen by many as further evidence for the string theory landscape and the application of the anthropic principle.

The inflation theory says that our universe began on a hill (or ledge) of potential vacuum energies. The universe began to roll down that hill rapidly — that is, our universe began expanding at an exponential rate — until we settled into a valley of vacuum energy. The question that eternal inflation tries to answer is: *Why did we start on that hill?*

Seemingly, the universe started with a random starting point on the spectrum of possible energies, so it's only luck that we were on the hill and, in turn, luck that we went through the right amount of inflation to distribute mass and energy the way it's distributed.

Or, alternately, there are a vast number of possibilities, many of which spring into existence, and we could only possibly exist in the ones that have this specific starting condition. (This is, in essence, the anthropic argument.)

In either case, the particles and forces of our universe are determined by the initial location on that hill and the laws of physics that govern how the universe will change over time.

In 1977, Sidney Coleman and Frank De Luccia described how quantum fluctuations in an inflating universe create tiny bubbles in the fabric of space-time. These bubbles can be treated as small universes in their own right (see Chapter 15). For now, the key is that they do form.

The cosmologist Andrei Linde has been the one to most strenuously argue that this finding, in combination with Alan Guth's inflationary theory, demands eternal inflation — the creation of a vast population of universes, each with slightly different physical properties. He has been joined by Guth himself and Alexander Vilenkin, who helped hammer out the key aspects of the theory.

The eternal inflation model says that these *bubble universes* (Guth prefers "pocket universes," while Susskind calls them "island universes") spring up, somehow getting physical laws among the possible ones dictated by the string theory landscape (through some as-yet-unknown means). The bubble universe then undergoes inflation. Meanwhile, the space around it continues to expand — and it expands so quickly that information about the inflating bubble universe can never reach another universe. Our own universe is one of these bubble universes, but one which finished its inflationary period long ago.

The hidden matter and energy

Two mysteries of our universe are the dark matter and dark energy (Chapter 9 contains the basics about these concepts). Dark matter is unseen matter that holds stars together in galaxies, while dark energy is unseen vacuum energy that pushes different galaxies farther apart from each other. String theory holds several possibilities for both.

A stringy look at dark matter

String theory provides a natural candidate for dark matter in supersymmetric particles, which are needed to make the theory work but which scientists have never observed. Alternatively, it's possible that dark matter somehow results from the gravitational influence of nearby branes.

Probably the simplest explanation of dark matter would be a vast sea of supersymmetric particles residing inside galaxies, but we can't see them (presumably because of some unknown properties of these new particles). Supersymmetry implies that every particle science knows about has a super-partner (see Chapter 10 if you need a refresher on supersymmetry). Fermions have bosonic superpartners and bosons have fermionic superpartners. In fact, one popular candidate for the missing dark matter is the photino, the superpartner of the photon.

A computer simulation, reported in the journal *Nature* in November 2008, offers one possible means of testing this idea. The simulation, performed by the international Virgo Consortium research group, suggests that dark matter in the halo of the Milky Way galaxy should produce detectable levels of gamma rays. This simulation indicates a direction to start looking for such tell-tale signs, at least.

Another possible dark matter candidate comes from the various brane world scenarios. Though the details still have to be worked out, it's possible that there are branes that overlap with our own 3-brane. Perhaps where we have galaxies, there are gravitational objects that extend into other branes. Because gravity is the one force that can interact across the branes, it's possible that these hyper-dimensional objects create added gravity within our own 3-brane.

Finally, the 4-dimensional string theories discussed in Chapter 13 present yet another possibility, because they require not only supersymmetry but a vast number of families of particles beyond the electron, muon, and tau families in our current Standard Model. Bringing string theory down to four dimensions seems to greatly expand the number of particles that physicists would expect to find in the universe, and (if they exist) these could account for dark matter.

A stringy look at dark energy

Even more intriguing than dark matter is dark energy, which is a positive energy that seems to permeate the entire universe and to be much more abundant than either form of matter — but also much less abundant than physicists think it *should* be. Recent discoveries in string theory have allowed for this dark energy to exist with in the theory.

Although string theory offers some possibilities for dark matter, it offers less explanation for dark energy. Theoretically, dark energy should be explained by the value of the vacuum energy in particle physics, where particles are continually created and destroyed. These quantum fluctuations grow immensely, leading to infinite values. (I explain in Chapter 8 that to avoid these infinite values in quantum field theory, the process of renormalization is used, which is essentially rounding the quantity to a noninfinite value. This wouldn't be viewed as a favorable method, except for the fact that it works.)

However, when physicists try to use their standard methods to compute the value of the vacuum energy, they get a value that is off from the experimental value of dark energy by 10^{120}!

The real value is incredibly small, but not quite zero. Though the amount of dark energy in the universe is vast (according to recent data, it makes up about 73 percent of the universe), the intensity of dark energy is very small — so small that until 1998, scientists assumed the value was exactly zero.

The existence of dark energy (or a positive cosmological constant, depending on how you want to look at it) doesn't remove the many solutions of string theory relating to different possible physical laws. The number of solutions that include dark energy may be on the order of 10^{500}. This dark energy reflects a positive energy built into the very fabric of the universe, likely related to the energy of the vacuum itself.

To some, the ekpyrotic universe has a benefit over the inflationary model, because it offers a reason for why we might observe such a value for dark energy in our universe: That's the part of the cyclic phase that we're in. At times in the past, the dark energy may have been stronger, and at times in the future it may be less. To many others, this reason isn't any more intellectually satisfying than the lack of a reason in other cosmological models. It still amounts to an accidental coincidence (or an application of the anthropic principle, as discussed later in this chapter).

Outside of the ekpyrotic universe, there's little explanation for what's going on. The problem of offsetting the expected vacuum energy by such a large amount — enough to *almost,* but not quite, cancel it out — is seen by many physicists as too much chance to contemplate.

Many would rather turn to the anthropic principle to explain it. Others see that as waving a white flag of surrender, admitting that dark energy is just too tough of a challenge to figure out.

The Undiscovered Country: The Future of the Cosmos

In cosmology, the past and the future are linked together, and the explanation for one is tied to the explanation of the other. With the big bang model in place, there are essentially three possible futures for our universe. Determining the solutions to string theory that apply to our universe might allow us to determine which future is most likely.

A universe of ice: The big freeze

In this model of the universe's future, the universe continues to expand forever. Energy slowly dissipates across a wider and wider volume of space and, eventually, the result is a vast cold expanse of space as the stars die. This *big freeze* has always had some degree of popularity, dating back to the rise of thermodynamics in the 1800s.

The laws of thermodynamics tell you that the entropy, or disorder, in a system will always increase. This means that the heat will spread out. In the context of cosmology, this means that the stars will die and their energy will radiate outward. In this "heat death," the universe becomes a static soup of near–absolute zero energy. The universe as a whole reaches a state of thermal equilibrium, meaning that nothing interesting can really happen.

A slightly different version of the big freeze model is based on the more recent discovery of dark energy. In this case, the repulsive gravity of dark energy will cause clusters of a galaxy to move apart from each other, while, on the smaller scale, those clusters will gather closer together, eventually forming one large galaxy.

Over time, the universe will be populated by large galaxies that are extremely far apart from each other. The galaxies will become inhospitable to life, and the other galaxies will be too far away to even see. This variant, sometimes called a "cold death," is another way the universe could end as a frozen wasteland. (This timescale is incredibly vast, and humans will likely not even still exist. So no need to panic.)

From point to point: The big crunch

One model for the future of the universe is that the mass density of the universe is high enough that the attractive gravity will eventually overpower the repulsive gravity of dark energy. In this *big crunch* model, the universe contracts back into a microscopic point of mass.

This idea of a big crunch was a popular notion when I was in high school and reading science fiction, but with the discovery of the repulsive dark energy, it seems to have gone out of favor. Because physicists are observing the expansion rate increase, it's unlikely that there's enough matter to overcome that and pull it all back together.

A new beginning: The big bounce

The ekpyrotic model (see the earlier section "A brane-fueled, 21st-century cyclic model: The ekpyrotic universe") brings the big crunch back, but with a twist. When the crunch occurs, the universe once again goes through a big bang period. This is not the only model that allows for such a *big bounce* cyclic model.

In the ekpyrotic model, the universe goes through a series of big bangs, followed by expansion and then a contracting big crunch. The cycle repeats over and over, presumably without any beginning or end. Cyclic models of the universe are not original, going back not only to 1930s physics, but also to religions, such as some interpretations of Hinduism.

It turns out that string theory's major competitor — loop quantum gravity (explained in Chapter 18) — may also present a big bounce picture. The method of loop quantum gravity is to *quantize* (break up into discrete units) space-time itself, and this avoids a singularity at the formation of the universe, which means that it's possible that time extends back before the big bang moment. In such a picture, a big bounce scenario is likely.

Exploring a Finely Tuned Universe

One major issue in cosmology for years has been the apparent fine-tuning seen in our universe. The universe seems specially crafted to allow life. One of the major explanations for this is the anthropic principle, which many string theorists have recently begun adopting. Many physicists still feel that the anthropic principle is a poor substitute for an explanation of why these physical properties must have the values they do.

To a physicist, the universe looks as if it were made for the creation of life. British Astronomer Royal Martin Rees clearly illuminated this situation in his 1999 book *Just Six Numbers: The Deep Forces That Shape the Universe*. In this book, Rees points out that there are many values — the intensity of dark energy, gravity, electromagnetic forces, atomic binding energies, to name just a few — that would, if different by even an extremely small amount, result in a universe that is inhospitable to life as we know it. (In some cases, the universe would have collapsed only moments after creation, resulting in a universe inhospitable for *any* form of life.)

The goal of science has always been to explain why nature has to have these values. This idea was once posed by Einstein's famous question: *Did God have a choice in creating the universe?*

Einstein's religious views are complex, but what he meant by this question wasn't actually so much religious as scientific. In other words, he was wondering if there was a fundamental reason — buried in the laws of nature — why the universe turned out the way it did.

For years, scientists sought to explain the way the universe worked in terms of fundamental principles that dictate the way the universe has formed. However, with string theory (and eternal inflation), that very process has resulted in answers that imply the existence of a vast number of universes and a vast number of scientific laws, which could be applied in those universes.

The major success of the anthropic principle is that it provided one of the only predictions for a small, but positive, cosmological constant prior to the discovery of dark energy. This was put forward in the 1986 book *The Anthropic Cosmological Principle* by John D. Barrow and Frank J. Tipler, and cosmologists in the 1980s appeared to be at least open-minded about the possibility of using anthropic reasoning.

Nobel laureate Steven Weinberg made the big case for anthropic reasoning in 1987. Analyzing details of how the universe formed, he realized two things:

- ✔ If the cosmological constant were negative, the universe would quickly collapse.
- ✔ If the cosmological constant were slightly larger than the experimentally possible value, matter would have been pushed apart too quickly for galaxies to form.

In other words, Weinberg realized that if scientists based their analysis on what was required to make life possible, then the cosmological constant couldn't be negative and had to be very small. There was no reason, in his analysis, for it to be exactly zero. A little over a decade later, astronomers discovered dark energy, which fit the cosmological constant in precisely the range specified by Weinberg. Martin Rees appealed to this type of discovery in his explanation of how the laws in our universe end up with such finely tuned values, including the cosmological constant.

You may wonder if there's anything particularly anthropic about Weinberg's reasoning, however. You only have to look around to realize that the universe didn't collapse and galaxies were able to form. It seems like this argument could be made just by observation.

The problem is that physicists are looking not only to determine the properties of our universe, but to explain them. To use this reasoning to explain the special status of our universe (that is, it contains us) requires something very important — a large number of *other* universes, most of which have properties that make them significantly different from us.

For an analogy, consider you're driving along and get a flat tire. If you were the only person who had ever gotten a flat tire, you might be tempted to explain the reason why you, out of everyone on the planet, were the one to get the flat tire. Knowing that many people get flat tires every day, no further explanation is needed — you just happen to have been in one of many cars that happened to get a flat tire.

If there is only one universe, then having the fine-tuned numbers that Rees and others note is a miraculously fortunate turn of events. If there are billions of universes, each with random laws from hundreds of billions (or more) possible laws from the string theory landscape, then every once in a while a universe like ours will be created. No further explanation is necessary.

The problem with the anthropic principle is that it tends to be a last resort for physicists. Scientists only turn to the anthropic principle when more conventional methods of arguments have failed them, and the second they can come up with a different explanation, they abandon it.

This isn't to imply that the scientists applying the anthropic principle are anything but sincere. Those who adopt it seem to believe that the vast string theory landscape — realized in a multiverse of possible universes (see Chapter 15) — can be used to explain the properties of our universe.

Chapter 15

Parallel Universes: Maybe You Can Be Two Places at Once

....

In This Chapter

▶ Examining the four types of parallel universes

▶ Using holes and tunnels to check out other universes

▶ Explaining our universe by our presence

....

String theory and its infant sibling, string cosmology, certainly give us amazing possibilities for what could be out there in our universe, but they also give us even more amazing possibilities about what could be out there *beyond* our universe, in other universes that may or may not have any connection with ours.

In this chapter, I explain what science in general, and string theory in particular, has to tell us about the possible existence of alternate universes. I start with a general discussion of these different types of parallel universes and then get into the specific traits of each. I also take a brief look at how quantum physics could possibly provide a way for intelligent beings from one universe to possibly contact another universe. Finally, the anthropic principle comes up again, and I explain how it relates to the ideas of parallel universes.

Exploring the Multiverse: A Theory of Parallel Universes

The *multiverse* is a theory in which our universe is not the only one, but states that many universes exist parallel to each other. These distinct universes within the multiverse theory are called *parallel universes*. A variety of different theories lend themselves to a multiverse viewpoint.

Multiverses in religion and philosophy

The idea of a physical multiverse came later to physics than in some other areas. The Hindu religion has ancient concepts that are similar. The term itself was, apparently, first applied by a psychologist, rather than a physicist.

Concepts of a multiverse are evident in the cyclical infinite worlds of ancient Hindu cosmology. In this viewpoint, our world is one of an infinite number of distinct worlds, each governed by its own gods on their own cycles of creation and destruction.

The word *multiverse* was originated by American psychologist William James in 1895

(the word "moral" is excluded from some citations of this passage):

> "Visible nature is all plasticity and indifference, a [moral] multiverse, as one might call it, and not a [moral] universe."

The phrase rose in prominence throughout the 20th century, when it was used regularly in science fiction and fantasy, notably in the work of author Michael Moorcock (though some sources attribute the word to the earlier work of author and philosopher John Cowper Powys in the 1950s). It is now a common phrase within these genres.

In some theories, there are copies of you sitting right here right now reading this book in other universes and other copies of you that are doing other things in other universes. Other theories contain parallel universes that are so radically different from our own that they follow entirely different fundamental laws of physics (or at least the same laws manifest in fundamentally different ways), likely collapsing or expanding so quickly that life never develops.

Not all physicists really believe that these universes exist. Even fewer believe that it would ever be possible to contact these parallel universes, likely not even in the entire span of our universe's history. Others believe the quantum physics adage that if it's possible, it's bound to happen somewhere and sometime, meaning it may be inevitable that quantum effects allow contact between parallel universes.

According to MIT cosmologist Max Tegmark, there are four levels of parallel universes:

✔ **Level 1:** An infinite universe that, by the laws of probability, must contain another copy of Earth somewhere

✔ **Level 2:** Other distant regions of space with different physical parameters, but the same basic laws

✔ **Level 3:** Other universes where each possibility that can exist does exist, as described by the many worlds interpretation (MWI) of quantum physics

✓ **Level 4:** Entirely distinct universes that may not even be connected to ours in any meaningful way and very likely have entirely different fundamental physical laws

The following sections look at each of these levels in more detail.

Tegmark's approach is one of the only attempts to comprehensively categorize the concepts of parallel universes in a scientific (or, as some see it, pseudoscientific) context. The full text of Tegmark's 2003 paper on this topic is available at his MIT Web site, space.mit.edu/home/tegmark/multiverse.pdf, for those who don't believe that these concepts are scientific. (They may not be scientific, but if that's the case, then at least they're unscientific musings by a scientist.)

Plurality of worlds: A hot topic

Early astronomy provided some support for the existence of a *plurality of worlds,* a view that was so controversial that it contributed to at least one man's death. These plurality of worlds, and the eventual parallel worlds, were rooted in the ideas of an infinite universe, as are the ideas of parallel universes presented in this chapter.

The Italian philosopher Giordano Bruno (1548–1600) was executed for a variety of heresies against the Catholic Church. Though Bruno was a supporter of the Copernican system, his abnormal beliefs went far beyond that: He believed in an eternal and infinite universe that contained a plurality of worlds. Bruno reasoned that because God was infinite, his creation would similarly be infinite. Each star was another sun, like our own, about which other worlds revolved. He didn't feel that such viewpoints were in opposition to the scriptures.

In fairness to the Catholic Church, Bruno wasn't executed merely for believing in other worlds. His list of heresies was long and varied, including denial of Mary's virginity, the divinity of Christ, the Trinity, the Incarnation, and the Catholic doctrine of transubstantiation. He also believed in reincarnation and was accused of practicing magic. This is not to say that any (or

all) of these viewpoints warranted death, but given the time period, it would be hard to get out of such accusations alive.

In 1686, the French writer Bernard le Bovier de Fontenelle wrote *Conversations on the Plurality of Worlds,* which was one of the first books to address the popular audience on scientific topics, being written in French rather than scholarly Latin. In *Conversations,* he explained the Copernican heliocentric model of the universe and contemplated extraterrestrial life. Though other enlightenment thinkers — possibly even John Adams and Benjamin Franklin, by some accounts — were agreeable to such viewpoints, it would be many years before the plurality of worlds extended to the plurality of universes.

In 1871, the French political malcontent Louis Auguste Blanqui wrote — while in prison — a brochure titled *Eternity by the Stars: Astronomical Hypotheses,* in which he said that an infinite universe would have to replicate the original set of combinations an infinite number of times to fill up the infinite space. This is, to my knowledge, the first inkling of the transition from "plurality of worlds" to "parallel worlds" — copies of you sitting reading this same book on another planet.

Level 1: If you go far enough, you'll get back home

The idea of Level 1 parallel universes basically says that space is so big that the rules of probability imply that surely, somewhere else out there, are other planets exactly like Earth. In fact, an infinite universe would have infinitely many planets, and on some of them, the events that play out would be virtually identical to those on our own Earth.

We don't see these other universes because our cosmic vision is limited by the speed of light — the ultimate speed limit. Light started traveling at the moment of the big bang, about 14 billion years ago, and so we can't see any further than about 14 billion light-years (a bit farther, since space is expanding). This volume of space is called the *Hubble volume* and represents our observable universe.

The existence of Level 1 parallel universes depends on two assumptions:

- ✔ The universe is infinite (or virtually so).
- ✔ Within an infinite universe, every single possible configuration of particles in a Hubble volume takes place multiple times.

In regard to the first assumption, inflation theory predicts that the universe is actually far larger than our Hubble volume. Recall that eternal inflation implies that universes are constantly being created and destroyed by quantum fluctuations, which means that space is actually infinite under the most extreme application of this theory.

The regions created in an eternal inflation model trigger every single set of initial conditions, leading to the second assumption. This means that there's another region of space that consists of a Hubble volume that has the exact same initial conditions as our universe. If it has exactly the same initial conditions, then such a region would evolve into a Hubble volume that resembles ours exactly.

If Level 1 parallel universes do exist, reaching one is virtually (but not entirely) impossible. For one thing, we wouldn't know where to look for one because, by definition, a Level 1 parallel universe is so far away that no message can ever get from us to them, or them to us. (Remember, we can only get messages from within our own Hubble volume.)

In theory, however, you could get in a spaceship that can travel at nearly the speed of light, point it in a direction, and head off. Time for you would slow, but the universe would continue to age as you moved throughout the entire expanse of the universe looking for your twin. If you're lucky, and dark energy is weak enough that eventually gravity causes cosmic expansion to end, you might eventually be able to get to your twin's planet.

Chaotic and eternal: Two facets of inflation

The theories of eternal inflation and chaotic inflation can be quite confusing, as I discovered in writing this book. Most people, even physicists, use them fairly interchangeably. This is an excellent example of how concepts on the cutting edge of science can get blurred, even between different experts in the field.

In eternal inflation, the quantum fluctuations in the vacuum energy result in "bubble universes" (or "pocket universes" or "island universes" . . . will the naming confusion never cease?!). The possible energies that such a universe could have (called the *false vacuum*) are represented by a graph that looks kind of like a mountain range, which is often referred to as an *energy hill*. The *true vacuum* of our universe is represented as one of the valleys in such a graph.

In 1983, Paul Steinhardt and Alex Vilenkin both presented the key ideas of eternal inflation, which is that quantum fluctuations can cause the triggering of new inflationary cycles. The assumption at the time was that each new cycle of inflation would start at the top of the energy hill and, during the inflationary cycle, would progress down toward the true vacuum. The energy state of the universe is decaying into a ground state.

In 1986, Andrei Linde wrote a paper called "Chaotic Inflation," in which he pointed out that these universes can be created anywhere on the energy hill, not necessarily at the peak. In fact, the hill itself may not even have a peak; it might continue on forever! He furthermore showed that chaotic inflation is also eternal, because it spawns continued creation of new bubble universes.

Several sources make chaotic inflation sound like a specific type of eternal inflation theory. Max Tegmark's 2003 article uses "chaotic inflation" in a way that sounds, to me, more like eternal inflation. Wikipedia has an article on chaotic inflation, identifying it as a "sub-class of eternal inflation," but has no article on eternal inflation itself!

But Vilenkin, in his 2006 book, *Many Worlds in One: The Search for Other Universes,* is adamant that chaotic inflation is an entirely different theory, seeming a bit frustrated that they're so often interchanged, a frustration that certainly seems justified, unless Vilenkin is the one who's applying the term imprecisely.

Time will tell what consensus cosmologists reach over this distinction between chaotic inflation and eternal inflation. For now, though, it's useful to know that most chaotic models will yield eternal inflation (but not all of them), and many eternal inflation models are not chaotic.

Level 2: If you go far enough, you'll fall into wonderland

In a Level 2 parallel universe, regions of space are continuing to undergo an inflation phase. Because of the continuing inflationary phase in these universes, space between us and the other universes is literally expanding faster than the speed of light — and they are, therefore, completely unreachable.

Two possible theories present reasons to believe that Level 2 parallel universes may exist: eternal inflation and ekpyrotic theory. Both theories were introduced in Chapter 14, but now you can see one of the consequences in action.

In eternal inflation, recall that the quantum fluctuations in the early universe's vacuum energy caused bubble universes to be created all over the place, expanding through their inflation stages at different rates. The initial condition of these universes is assumed to be at a maximum energy level, although at least one variant, *chaotic inflation,* predicts that the initial condition can be chaotically chosen as any energy level, which may have no maximum, and the results will be the same. (See the nearby sidebar "Chaotic and eternal: Two facets of inflation" for more information.)

The findings of eternal inflation mean that when inflation starts, it produces not just one universe, but an infinite number of universes.

Right now, the only noninflationary model that carries any kind of weight is the ekpyrotic model, which is so new that it's still highly speculative. (Ironically, both the eternal inflation model and the ekpyrotic model were partially created by cosmologist Paul Steinhardt.)

In the ekpyrotic theory picture, if the universe is the region that results when two branes collide, then the branes could actually collide in multiple locations. Consider flapping a sheet up and down rapidly onto the surface of a bed. The sheet doesn't touch the bed only in one location, but rather touches it in multiple locations. If the sheet were a brane, then each point of collision would create its own universe with its own initial conditions.

There's no reason to expect that branes collide in only one place, so the ekpyrotic theory makes it very probable that there are other universes in other locations, expanding even as you consider this possibility.

In other words, modern cosmology — regardless of whether inflation or ekpyrosis are true — virtually demands that Level 2 parallel universes exist. (Some alternate cosmological theories presented in Chapter 19, such as variable speed of light cosmology and modified gravity, don't have this demand.)

As in the Level 1 universes, these universes would be created with essentially random initial conditions, which, averaged out over infinity, implies that there are other universes that are virtually (or completely) identical to our own. These new universes are continually formed, so many (infinitely many, in fact) are still undergoing the inflationary phase of their evolution.

Unlike in a Level 1 universe, it's possible that a Level 2 universe could have different fundamental properties, such as a higher (or lower) number of dimensions, a different array of elementary particles, fundamental force strengths, and so on. But these universes are created by the same laws of physics that created ours, just with different parameters. These universes could behave quite differently from our own, but the laws that govern them would — on a very fundamental level — be exactly the same.

Unfortunately, Level 2 universes are pretty much impossible to reach. Not only are there an infinite number of universes, but there are an infinite number of inflating universes, which means the space between our universe and a parallel universe is expanding. So even if we could move at the speed of light (and we can't), we'd never be able to get to another universe. Space itself is inflating faster than we can move between our universe and another Level 2 universe.

Level 3: If you stay where you are, you'll run into yourself

A Level 3 parallel universe is a consequence of the many worlds interpretation (MWI) from quantum physics. In this interpretation, every single quantum possibility inherent in the quantum wavefunction becomes a real possibility in some reality. When the average person (especially a science fiction fan) thinks of a "parallel universe," he's probably thinking of Level 3 parallel universes.

The many worlds interpretation was presented by Hugh Everett III to explain the quantum wavefunction, the Schrödinger equation. The Schrödinger equation describes how a quantum system evolves over time through a series of rotations in a *Hilbert space* (an abstract space with infinite dimensions). The evolution of the wavefunction is called unitary. (*Unitarity* basically means that if you add up the probabilities of all possible outcomes, you get 1 as the sum of those probabilities.)

The traditional Copenhagen interpretation of quantum physics assumed that the wavefunction collapsed into a specific state, but the theory presented no mechanism for when or how this collapse takes place. The collapse turned the unitary wavefunction, which contains all possibilities, into a non-unitary system, which ignores the possibilities that never took place.

Everett took a tactic similar to that taken later by string theorists, assuming that each "dimension" predicted mathematically by the wavefunction (an infinite number of them) must be realized in some way in reality. In this theory, all quantum events result in a branching of a universe into multiple universes, so the unitary theory can be treated in a unitary way (no possibilities ever go away).

Alternate history across many worlds

Of all the types of parallel universes, Level 3 universes have most captured the imagination of popular culture, spawning their own genre of science fiction and fantasy: *alternate history*. These are stories written with settings that are based on our own universe, but with the assumption that some historical event went differently, resulting in consequences different from those in our own universe. (For the non-science fiction fan, think of *It's a Wonderful Life*.) In these fictional universes, it's possible (and common) that visitors from one universe can interact with a Level 3 parallel universe.

Obviously, in these fictional universes, the author (and reader) care about the macroscopic differences, but the many worlds interpretation applies to all levels. If a particle decays, or not, different worlds represent those events. No one

observing would be able to tell the difference between them. However, if they were observing with a Geiger counter, which detects radioactive decay, the quantum split would result in further splits. The Geiger counter is triggered in one universe and not the other. The scientist who detects the decay would react differently, perhaps, than the one who does not detect the decay. So, in principle, this is how these tiny quantum universes become full-fledged parallel universes.

In fiction, the effects are generally more dramatic, such as the southern states winning the American Civil War or the Byzantine Empire never collapsing (both of which have been explored by alternate history author Harry Turtledove, called "the Master of Alternate History" by his fans).

Level 3 parallel universes are different from the others posed because they take place in the same space and time as our own universe, but you still have no way to access them. You have never had and will never have contact with any Level 1 or Level 2 universe (I assume), but you're continually in contact with Level 3 universes — every moment of your life, every decision you make, is causing a split of your "now" self into an infinite number of future selves, all of which are unaware of each other.

Though we talk of the universe "splitting," this isn't precisely true (under the MWI of quantum physics). From a mathematical standpoint, there's only one wavefunction, and it evolves over time. The *superpositions* of different universes all coexist simultaneously in the same infinite-dimensional Hilbert space. These separate, coexisting universes interfere with each other, yielding the bizarre quantum behaviors, such as those of the double slit experiment in Chapter 7.

Of the four types of universes, Level 3 parallel universes have the least to do with string theory directly.

Level 4: Somewhere over the rainbow, there's a magical land

A Level 4 parallel universe is the strangest place (and most controversial prediction) of all, because it would follow fundamentally different mathematical laws of nature than our universe. In short, any universe that physicists can get to work out on paper would exist, based on the *mathematical democracy principle:* Any universe that is mathematically possible has equal possibility of actually existing.

Scientists use mathematics as their primary tool to express the theories of how nature behaves. In a sense, the mathematics that represents the theory is the meat of the theory, the thing that really gives it substance.

In 1960, physicist Eugene Wigner published an article with the provocative title "The Unreasonable Effectiveness of Mathematics in the Natural Sciences," in which he pointed out that it's kind of unreasonable that mathematics — a construct purely of the mind — would be so good at describing physical laws. He went further than this, supposing that this effectiveness represented a deep level of connection between mathematics and physics, and that by exploring mathematics you can figure out ways to approach sciences in new and innovative ways.

But the equations that work so well to describe our universe are only one set of equations. Certainly a universe could be created, as physicists have done on paper, with only two dimensions and containing no matter, which is nothing but expanding space. There could be a vast anti-de Sitter space, contracting, right next to it.

Why, then, do we observe the specific set of equations, specific set of laws, that we do? In other words, to use the phrase of British cosmologist Stephen Hawking (from his 1988 *A Brief History of Time*), what is the force that "breathes fire into the equations" that govern our universe?

Throughout this book, you explore concepts that are on the cutting edge of theoretical physics — the bosonic string theory, the various superstring theories, AdS/CFT correspondence, Randall-Sundrum models — but that clearly don't match our own universe. Most physicists leave it at that, with the understanding that some "pure math" just doesn't apply directly to the physical universe we live in. However, according to the principle of mathematical democracy, *these universes do exist somewhere.*

In this controversial view, our equations aren't preferred, but in the multi-verse, every equation that can have life breathed into it will. This makes up the Level 4 multiverse, a place so vast and strange that even the most brilliant among us can only conceptualize it with the tools of mathematics.

Accessing Other Universes

With the four types of parallel universe described, it's time to look at the fun part — whether there's any way to reach them. Realistically, the answer is probably "No," but that's not the most interesting option, so the following sections look into ways that some of these universes might be able to inter-act with our own.

A history of hyperspace

To access a Level 1, 2, or 4 universe, you'd have to find a way to traverse an incredibly large distance in a moment's time, a task made more difficult by Einstein's speed limit — the speed of light. One of the only ways to achieve this would be by using extra space dimensions — sometimes called *hyperspace* — to cut down the distance.

Where are those extra dimensions, anyway?

Current string theory models postulate ten space dimensions (plus one time dimension). Our observed universe appears to have only three space dimen-sions (plus the one time dimension). String theory offers two possibilities for the extra dimensions:

- ✔ The extra seven dimensions extend off of a 3-brane on which our uni-verse resides.

- ✔ The extra seven dimensions are compactified (likely into a Planck length radius shape), while our three space dimensions are uncompactified. (This is the dominant string theory viewpoint.)

You can picture a modified version of the first possibility by looking at Figure 15-1, which shows a universe of people living on a 2-brane. A third dimension extends off of that brane.

In theory, there could be some means for the 2-brane residents to leave the 2-brane and experience the greater 3-dimensional reality, as in the classic novel *Flatland*. By extension, there could be a way for people in our universe to leave our 3-brane to travel in the extra dimensions.

Figure 15-1:
It's possible that there are dimensions beyond our own, which we can't access.

For the second possibility, the dimensions are compactified to sizes that are so small no one has ever observed them. As discussed in Chapter 11, some recent theories have indicated that these sizes could be as large as a visible fraction of a millimeter, and tests along these lines should take place at the Large Hadron Collider. Some speculative ideas (not even well developed enough to be called theories) have been put forth that these compactified dimensions could contain their own universes.

String theory also allows for the possibility that some regions of the universe would have large extra dimensions, allowing them to interact with the current three dimensions in meaningful ways. No models suggest that this is actually happening in our universe, but the theory allows for such behavior.

Wormholes: Busting out of three-space

Even before string theory, the idea existed that the geometry of the universe would allow for shorter paths between points. In fiction, this can be seen in stories such as Lewis Carroll's _Alice in Wonderland,_ and in science, it can be seen in the wormholes, as depicted in Figure 1-4 in Chapter 1.

A _wormhole_ is a shortcut to go from one location on a surface to another, just as a worm can dig through the center of an apple to get from one surface to another (thus the name). This concept arises from Einstein's theory of general relativity, proposed years before string theory was conceived. These traditional wormholes connect different regions in the same universe and, as you can see in Chapter 16, have been exploited for many outlandish theoretical purposes, despite the fact that no one knows for sure whether they exist. (So what's one more!)

Similarly, it's possible that in a brane world scenario, we are somehow touching or connected to another brane. If these branes overlap, it's conceivable that there would be a way to travel from the space of one brane to the space of another brane. (This is not the standard way that multiple branes interact in string theory. Much more common are the brane world scenarios from Chapter 11, where separate branes host difference pieces of the physics of our universe and then interact gravitationally.)

It's unlikely that such a brane jump would take place merely by stepping into a mirror, but something as powerful as a wormhole might do the trick. It's possible that a wormhole — generally hypothesized by general relativity as existing within rotating black holes and being notoriously unstable — might allow bits of matter or energy to slip from one universe into another parallel universe. If such strange events occurred at points where different branes overlapped in the *bulk* (the greater space that contains all of the branes), it's unclear whether they might provide a way to get matter and energy from one brane to another as well.

In fact, one possible resolution to the black hole information paradox that has long been considered by some is the idea that information that enters a black hole exits into a parallel universe by means of a wormhole at the center of the black hole.

Such ideas are obviously highly speculative, but mathematical models have shown it's feasible that some sort of wormhole — if held open by a form of negative energy — could provide a means of connecting different portions of space.

If this is the case, then the arguments in favor of parallel universes are on our side, because given an infinite universe and infinite time, everything is bound to happen somewhere. In a universe where parallel universes exist, travel between them may be guaranteed.

How quantum mechanics can get us from here to there

One other process of getting from one universe to another would be to use the property of *quantum tunneling,* which is where a particle is allowed to "jump" from one location to another across a barrier.

As Chapter 7 reveals, the uncertainty principle of quantum physics means that particles don't have a definite location, but instead both the location and momentum of each particle are linked together with a sort of "fuzziness." The more precisely you determine the location, the more fuzzy the momentum is, and vice versa.

This principle results in a strange phenomenon, known as quantum tunneling and shown in Figure 15-2. In this case, there is some sort of barrier (usually a potential energy barrier) that the particle shouldn't be able to cross normally. But the graph, which represents the probability that the particle is in any given location, extends a bit across the barrier.

Figure 15-2:
According to quantum physics, sometimes particles can "tunnel" across barriers.

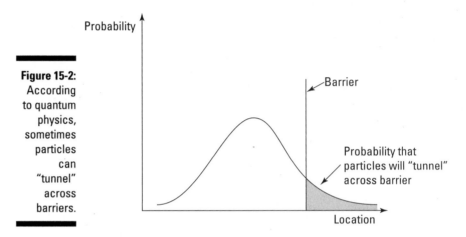

In other words, even when there's an uncrossable barrier, there's a *slight* chance — according to quantum mechanics — that a particle that should be on one side of the barrier may end up on the other side of the barrier. This behavior has been confirmed by experiment.

This provides a means that could *in theory* be used to access a parallel universe. Some cosmologists have suggested that exactly this physical mechanism is what started our own expansion as a universe.

The idea of quantum tunneling is key to the operation of electron-tunneling microscopes, which allow scientists to observe objects in incredibly fine detail.

Particles can only tunnel from a higher energy state into a lower energy state, though, so there are some limits on how this could be used, and the idea of using it to access another universe in a controlled way is way beyond current technology (or even current theory).

But for a sufficiently advanced civilization, one that has a theory that fully explains all aspects of physics and the ability to use vast amounts of energy, this sort of idea may be a possible means of getting to another universe.

Eaten by rogue universes

The assumption in this chapter has mostly been that the separate universes described don't normally interact with each other, but some approaches over the years have called this into question. One of the most recent is a 2008 paper in the journal *Physics Review D* by Eduardo Guendelman and Nobuyuki Sakai, in which they examine the idea of bubble universes to see if they could expand without the need for a big bang.

To make the equations work, Guendelman and Sakai had to introduce a repulsive *phantom energy,* which is possibly similar to dark energy. They found two types of stable solutions:

✔ The *child universe,* which is isolated from the parent universe (essentially a universe inside a black hole)

✔ A *rogue universe,* which is not isolated from the parent universe

This second kind of universe is troublesome, because as it begins to go through its inflation cycle, it does so by devouring the space-time of the parent universe. The parent universe is swept away as the rogue universe expands in its place — and it does so faster than the speed of light, so there's no warning.

Fortunately, there's no evidence that this phantom energy actually exists, or, if it does, it's possible that it exists in the form of dark energy (or inflation energy), which means that we may be one of these rogue universes ourselves. As our universe expands, it may be devouring some other, larger universe!

Chapter 16

Have Time, Will Travel

*O*ne of the most fascinating concepts in science fiction is the idea of traveling forward or backward in time, as in H. G. Wells's classic story *The Time Machine*. Scientists haven't been able to build a time machine yet, but some physicists believe that it may someday be possible — and some (probably most) believe that it will *never* be possible.

Time travel exists in physics because of possible solutions to Einstein's general theory of relativity, mostly resulting in singularities. These singularities would be eliminated by string theory, so in a universe where string theory dictates the laws of the universe, time travel will probably not be allowed — a result that many physicists find quite favorable to the alternative (though far less interesting).

In this chapter, I explore the notion of time and our travel through it — both in the normal, day-to-day method and in more unusual, speculative methods. I discuss the scientific meaning of time, in both classical terms and from the standpoint of special relativity. One possible method of time travel involves using cosmic strings. There's a possibility, which I explore, that there may be more than one time dimension. I also explain one scenario for creating a physically plausible (though probably impossible) time machine using wormholes. Finally, I look at some of the different logical paradoxes involved with time travel.

Temporal Mechanics 101: How Time Flies

We move through time every single day, and most of us don't even think about how fascinating it is. Scientists who have thought about it have constantly run into trouble in figuring out exactly what time means because time is such an abstract concept. It's something we're intimately familiar with, but so familiar with that we almost never have to analyze it in a meaningful way.

Over the years, our view of time — both individually and from a scientific standpoint — has changed dramatically, from an intuition about the passage of events to a fundamental component of the mathematical geometry that describes the universe.

The arrow of time: A one-way ticket

Physicists refer to the one-way motion through time (into the future and never the past) using the phrase "arrow of time," first used by Arthur Eddington (the guy who helped confirm general relativity) in his 1928 book *The Nature of the Physical World*. The first note he makes is that "time's arrow" points in one direction, as opposed to directions in space, where you can reorient as needed. He then points out three key ideas about the arrow of time:

- ✔ Human consciousness inherently recognizes the direction of time.
- ✔ Even without memory, the world only makes sense if the arrow of time points into the future.
- ✔ In physics, the only place the direction of time shows up is in the behavior of a large number of particles, in the form of the second law of thermodynamics. (See the nearby sidebar, "Time asymmetries," for a clarification of the exceptions to this.)

The conscious recognition of time is the first (and most significant) evidence that any of us has about the direction we travel in time. Our minds (along with the rest of us) "move" sequentially in one direction through time, and most definitely not in the other. The neural pathways form in our brain, which retains this record of events. In our minds, the past and future are distinctly different. The past is static and unchanging, but the future is fully undetermined (at least so far as our brain knows).

As Eddington pointed out, even if you didn't retain any sort of memory, logic would dictate that the past must have happened before the future. This is probably true, although whether one could conceptualize of a universe in which time flowed from the future to the past is a question that's open for debate.

Time asymmetries

Arthur Eddington's third observation about the arrow of time indicates that physical laws actually ignore the direction of time, except for the second law of thermodynamics. What this means is that if you take the time t in any physics equation and replace it with a time $-t$, and then perform the calculations to describe what takes place, you'll end up with a result that makes sense.

For gravity, electromagnetism, and the strong nuclear force, changing the sign on the time variable (called *T-symmetry*) allows the laws of physics to work perfectly well. In some special cases related to the weak nuclear force, this actually turns out not to be the case.

There is actually a larger type of symmetry, called *CPT symmetry,* which is always preserved. The C stands for *charge-conjugation symmetry,* which means that positive and negative charges switch. The P stands for *parity symmetry,* which involves basically replacing a particle for a complete mirror image — a particle that has been flipped across all three space dimensions. (This CPT symmetry is a property of quantum theory in our four-dimensional space-time, so at present we are ignoring the other six dimensions proposed by string theory.)

The total CPT symmetry, it turns out, appears to be preserved in nature. (This is one of the few cases of unbroken symmetry in our universe.) In other words, an *exact* mirror image of our universe — one with all matter swapped for antimatter, reflected in all spatial directions, and traveling backward in time — would obey physical laws that are identical to those of our own universe in every conceivable way.

If CP symmetry is violated, then there must be a corresponding break in T-symmetry, so the total CPT symmetry is preserved. In fact, the handful of processes that violate T-symmetry are called *CP violations* (because the CP violation is easier to test than a violation in the time-reversal symmetry).

Finally, though, we reach the physics of the situation: the second law of thermodynamics. According to this law, as time progresses, no *closed system* (that is, a system that isn't gaining energy from outside of the system) can lose *entropy* (disorder) as time progresses. In other words, as time goes on, it's not possible for a closed system to become more orderly.

Intuitively, this is certainly the case. If you look at a house that's been abandoned, it will grow disordered over time. For it to become more orderly, there has to be an introduction of work from outside the system. Someone has to mow the yard, clean the gutters, paint the walls, and so on. (This analogy isn't perfect, because even the abandoned house gets energy and influence from outside — sunlight, animals, rainfall, and so on — but you get the idea.)

In physics, the arrow in time is the direction in which entropy (disorder) increases. It's the direction of decay.

Oddly, these same ideas (the same in spirit, though not scientific) date all the way back to St. Augustine of Alexandria's *Confessions,* written in 400 BCE, where he said:

> "What then is time? If no one asks me, I know: if I wish to explain it to one that asketh, I know not: yet I say boldly that I know, that if nothing passed away, time were not; and if nothing were, time present were not."

What Augustine is pointing out here is the inherent problem in explaining the slippery nature of time. We know exactly what time is — in fact, we are unable *not* to understand how it flows in our own lives — but when we try to define it in precise terms, it eludes us. He speaks of "if nothing passed away, time were not" that could, in a sense, describe how the second law of thermodynamics defines time's arrow. We know time passes because things change in a certain way as time passes.

Relativity, worldlines, and worldsheets: Moving through space-time

Understanding how time travel works within string theory would require a complete understanding of how the fabric of space-time behaves within the theory. So far, string theory hasn't exactly figured that out.

In general relativity, the motion of objects through space-time is described by a worldline. In string theory, scientists talk about strings (and branes) creating entire *worldsheets* as they move through space-time.

Worldlines were originally constructed by Hermann Minkowski when he created his Minkowski diagrams, shown in Chapter 6. Similar diagrams return in the form of Feynman diagrams (see Chapter 8), which demonstrate the worldlines of particles as they interact with each other through the exchange of gauge bosons.

In string theory, instead of the straight worldlines of point particles, it is the movement of strings through space-time that interests scientists, as shown in the right side of Figure 16-1.

Notice that in the original Feynman diagram, shown on the left of Figure 16-1, there are sharp points where the worldlines intersect (representing the point where the particles interact). In the worldsheet, the virtual string exchanged between the two original strings creates a smooth curve that has no sharp points. This equates to the fact that string theory contains no infinities in the description of this interaction, as opposed to pure quantum field theory. (Removing the infinities in quantum field theory requires renormalization.)

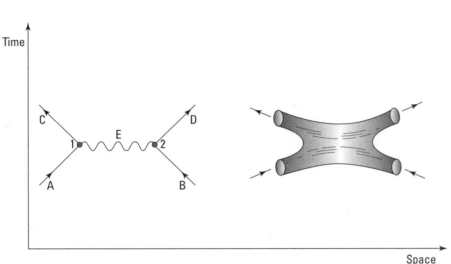

Figure 16-1: Instead of a world-line (left), a string creates a worldsheet (right) when it moves through space.

One problem with both quantum field theory and string theory is that they are constructed in a way that gets placed inside the space-time coordinate system. General relativity, on the other hand, depicts a universe in which the space-time is dynamic. String theorists hope string theory will solve this conflict between the background-dependent quantum field theory and the background-independent general relativity so that eventually dynamic space-time will be derived out of string theory. One criticism (as discussed in Chapter 17) is that string theory is, at present, still background-dependent.

The competing theory, loop quantum gravity, incorporates space into the theory, but is still mounted on a background of time coordinates. Loop quantum gravity is covered in more detail in Chapter 18.

Hawking's chronology protection conjecture: You're not going anywhere

The concept of time travel is often closely tied to infinities in the curvature of space-time, such as that within black holes. In fact, the discoveries of mathematically possible time travel were found in the general relativity equations containing extreme space-time curvature. Stephen Hawking, one of the most renowned experts in looking at space-time curvature, believes that time travel is impossible and has proposed a *chronology protection conjecture* that some mechanism must exist to prevent time travel.

When black holes were first proposed as solutions to Einstein's field equations, neither Einstein nor Eddington believed they were real. In a speech to a Royal Astronomical Society, Eddington said of black hole formation, "I think there should be a law of nature to prevent a star from behaving in this absurd way!"

Although Hawking is certainly comfortable with the idea of black holes, he objects to the idea of time travel. He proposed his chronology protection conjecture, which states that there must be something in the universe that prevents time travel.

Hawking's sometimes collaborator, Oxford physicist Roger Penrose, made the much more guarded claim that all singularities would be protected by an event horizon, which would shield them from direct interaction with our normal space-time, known as the *cosmic censorship conjecture.* This would also potentially prevent many forms of time travel from being accessible to the universe at large.

One major reason time travel causes so much trouble for physics (and must therefore be prohibited, according to Einstein and Hawking) is that you could create a way of generating an infinite amount of energy. Say you had a portal into the past and shone a laser into it. You set up mirrors so the light coming out of the portal is deflected back around to go into the portal again, in tandem with the original beam you have set up.

Now the total intensity of light coming out of the portal (in the past) would be (or have been) twice the original laser light going in. This laser light is sent back through the portal, yielding an output of four times as much light as originally transmitted. This process could be continued, resulting in literally an infinite amount of energy created instantaneously.

Obviously, such a situation is just one of many examples why physicists tend to doubt the possibility of time travel (with a few notable exceptions, which I cover throughout this chapter). If time travel were possible, then the predictive power of physics is lost, because the initial conditions are no longer trustworthy! The predictions based on those conditions would, therefore, be completely meaningless.

Slowing Time to a Standstill with Relativity

In physics, time travel is closely linked to Einstein's theory of relativity, which allows motion in space to actually alter the flow of time. This effect is known as *time dilation* and was one of the earliest predictions of relativity. This sort of time travel is completely allowed by the known laws of physics, but it allows only travel into the future, not into the past.

In this section I explore the special cases in relativity that imply that time travel — or at least altered motion through time — may in fact be possible. Skip ahead to "General Relativity and Wormholes: Doorways in Space and Time" for more information about how the general theory of relativity relates to potential time travel.

Time dilation and black hole event horizons, both of which I explain in the following sections, provide intriguing ways of extending human life, and in science fiction they've long provided the means for allowing humans to live long enough to travel from star to star. (See the later sidebar "The science fiction of time" for more information on this.)

Time dilation: Sometimes even the best watches run slow

The most evident case of time acting oddly in relativity, and one that has been experimentally verified, is the concept of time dilation under special relativity. *Time dilation* is the idea that as you move through space, time itself is measured differently for the moving object than the unmoving object. For motion that is near the speed of light, this effect is noticeable and allows a way to travel into the future faster than we normally do.

One experiment that confirms this strange behavior is based on unstable particles, pions and muons. Physicists know how quickly the particles would decay if they were sitting still, but when they bombard Earth in the form of cosmic rays, they're moving very quickly. Their decay rates don't match the predictions, but if you apply special relativity and consider the time from the particle's point of view, the time comes out as expected.

In fact, time dilation is confirmed by a number of experiments. In the Hafele-Keating experiments of 1971, atomic clocks (which are *very* precise) were flown on airplanes traveling in opposite directions. The time differences shown on the clocks, as a result of their relative motion, precisely matched the predictions from relativity. Also, global positioning system (GPS) satellites have to compensate for this time dilation to function properly. So time dilation is on very solid scientific ground.

Time dilation leads to one popular form of time travel. If you were to get into a spaceship that traveled very quickly away from Earth, time inside the ship would slow down in comparison to that on Earth. You could do a flyby of a nearby star and return to Earth at nearly the speed of light, and a few years would pass on Earth while possibly only a few weeks or months would pass for you, depending on how fast you were going and how far away the star was.

The biggest problem with this is how to accelerate a ship up to those speeds. Scientists and science fiction authors have made various proposals for such devices, but all are well outside the range of what we could feasibly build today or in the foreseeable future.

As you accelerate an object to high speeds, its mass also increases, which means it takes more and more energy to keep accelerating it. This formula of mass increase is similar to the formula that describes time dilation, so this makes it fundamentally difficult to get significant levels of time dilation.

The question is how much time dilation you really need, though, especially for trips within only a few light-years of Earth. One strange potential byproduct of this form of time travel is described toward the end of this chapter in the section entitled "The twin paradox."

Black hole event horizons: An extra-slow version of slow motion

One other case where time slows down, this time in general relativity, involves black holes. Recall that a black hole bends space-time itself, to the point where even light can't escape. This bending of space-time means that as you approach a black hole, time will slow down for you relative to the outside world.

If you were approaching the black hole and I were far away watching (and could somehow watch "instantly," without worrying about the time lag from light speed), I would see you approach the black hole, slow down and eventually come to rest to hover outside of it. Through the window of your spaceship, I would see you sitting absolutely still.

You, on the other hand, would not notice anything in particular — at least until the intense gravity of the black hole killed you. But until then, it certainly wouldn't "feel" like time was moving differently. You'd have no idea that as you glide past the black hole's event horizon (which you possibly wouldn't even notice), thousands of years were passing outside of the black hole.

As you find out in the next section, some believe that black holes may actually provide a means to more impressive forms of time travel as well.

General Relativity and Wormholes: Doorways in Space and Time

In general relativity, the fabric of space-time can occasionally allow for world-lines that create a *closed timelike curve,* which is relativity-speak for time travel. Einstein himself explored these concepts when developing general relativity, but never made much progress on them. In the following years, solutions allowing for time travel were discovered.

The first application of general relativity to time travel was by the Scottish physicist W. J. van Stockum in 1937. Van Stockum imagined (in mathematical form, because that's how physicists imagine things) an infinitely long, extremely dense rotating cylinder, like an endless barbershop pole. It turns out that in this case, the dense cylinder actually drags space-time with it, creating a space-time whirlpool.

This space-time whirlpool is an example of a phenomenon called *frame dragging*. It takes place when an object "drags" space (and time) along with it. This frame dragging is in addition to the normal bending of space-time due to gravity and is due to the movement of incredibly dense objects in space, such as neutron stars. This is similar to how an electric mixer causes the surrounding cake batter to swirl. This effect is frequently exploited to come up with time-travel solutions.

In van Stockum's situation, you could fly up to the cylinder in a spaceship and set a course around the cylinder and arrive back at a point in time before you arrived at the cylinder. In other words, you can travel into the past along a closed timelike curve. (If you can't picture this path, don't feel bad. The path is in four dimensions, after all, and results in going backward in time, so it's clearly something our brains didn't evolve to picture.)

Another theory about time travel was proposed in 1949 by Einstein's colleague and friend at Princeton University's Institute for Advanced Study, the mathematician Kurt Gödel. Gödel considered the situation where all of space — the entire universe itself — was actually rotating. You might ask if *everything* is rotating how we'd ever know it. Well, it turns out that if the universe were rotating, according to general relativity, then we'd see laser beams curve slightly as they move through space (beyond the normal gravitational lensing, where gravity bends light beams).

The solution that Gödel arrived at was disturbing, because it allowed time travel. It was possible to create a path in a rotating universe that ended before it began. In Gödel's rotating universe, the universe itself could function as a time machine.

So far, physicists haven't found any conclusive evidence that our universe is rotating. In fact, the evidence points overwhelmingly toward the idea that it's not. But even if the universe as a whole doesn't rotate, objects in it certainly do.

Taking a shortcut through space and time with a wormhole

In a solution called an *Einstein-Rosen bridge* (shown in Figure 16-2 and more commonly called a *wormhole*), two points in space-time could be connected by a shortened path. In some special cases, a wormhole may actually allow for time travel. Instead of connecting different regions of space, the wormhole could connect different regions of time!

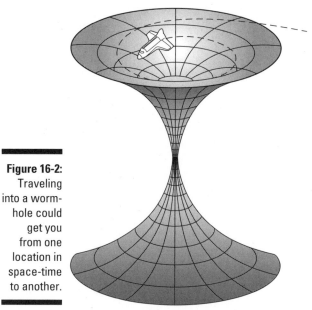

Figure 16-2:
Traveling
into a worm-
hole could
get you
from one
location in
space-time
to another.

Wormholes were studied by Albert Einstein and his pupil Nathan Rosen in 1935. (Ludwig Flamm had first proposed them in 1916.) In this model, the singularity at the center of a black hole is connected to another singularity, which results in a theoretical object called a *white hole*.

While the black hole draws matter into it, a white hole spits matter out. Mathematically, a white hole is a time-reversed black hole. Because no one's ever witnessed a white hole, it's probable that they don't exist, but they are allowed by the equations of general relativity and haven't been completely ruled out yet.

An object falling into a black hole could travel through the wormhole and come out the white hole on the other side in another region of space. Einstein showed that there were two flaws with using a wormhole for time travel:

✔ A wormhole is so unstable that it would collapse in upon itself almost instantaneously.

✔ Any object going into a black hole would be ripped apart by the intense gravitational force inside the black hole and would never make it out the other side.

Then, in 1963, New Zealand mathematician Roy Kerr calculated an exact solution for Einstein's field equations representing a *Kerr black hole*. The special feature of a Kerr black hole is that it rotates. So far as scientists know, *all* objects in the universe rotate, including stars, so when the star collapses into a black hole, it's likely that it too will rotate.

In Kerr's solution, it's actually possible to travel through the rotating black hole and miss the singularity at the center, so you could come out the other side. The problem is, again, that the black hole would probably collapse as you're going through it. (I address this problem in the next section.)

Assuming physicists could get a wormhole to be large and stable enough to pass through, probably the simplest time machine that could use this method was theorized by Kip Thorne of the California Institute of Technology. Consider a wormhole with the following features:

✔ One end of the wormhole is on Earth.

✔ The other end of the wormhole is located inside a spaceship, currently at rest on Earth. The end in the spaceship moves when the spaceship moves.

✔ You can travel through the wormhole either way, or talk through it, and such travel or communication is essentially instantaneous.

Now assume that a pair of twins, named Maggie and Emily, are standing at either end of the wormhole. Maggie is next to the wormhole on Earth in 2009, while Emily is on the spaceship (also, for the moment, in 2009). She goes on a little jaunt for a few days, traveling at nearly the speed of light, but when she comes back, thousands of years have passed on Earth due to time dilation (she is now in 5909).

On Maggie's side of the wormhole (still 2009), only a few days have passed. In fact, the twins have regularly been discussing the strange sights that Emily has witnessed over the few days of her journey. Emily (in 5909) is able to go through the wormhole to Maggie's location (in 2009) and, voilà, she has traveled back in time thousands of years!

In fact, now that Emily's gone to the trouble of setting up the portal, Maggie (or anyone else) could just as easily travel from 2009 to 5909 (or vice versa) just by stepping through it.

Since Thorne's model, there have been several wormhole-based time travel scenarios developed by physicists. In fact, some physicists have shown that if a wormhole exists, it *has* to allow travel in time as well as space.

Overcoming a wormhole's instability with negative energy

The problem with using wormholes to travel in space or time is that they are inherently unstable. When a particle enters a wormhole, it creates fluctuations that cause the structure to collapse in upon itself. There are theories that a wormhole could be held open by some form of *negative energy*, which represents a case where the energy density (energy per volume) of space is actually negative.

Under these theories, if a sufficient quantity of negative energy could be employed, it might continue to hold the wormhole open while objects pass through it. This would be an absolute necessity for any of the previously discussed theories that allow a wormhole to become a time portal, but scientists lack a real understanding of how to get enough negative energy together, and most think it's an impossible task.

In some models, it may be possible to relate dark energy and negative energy (both exhibit a form of repulsive gravity, even though dark energy is a positive energy), but these models are highly contrived. The good news (if you see possible time travel as good news) is that our universe appears to have dark energy in abundance, although the problem is that it looks like it's evenly distributed throughout the universe.

Trying to find any way to store negative energy and use it to sustain a wormhole's stability is far beyond current technology (if it's even possible at all). String theory can provide potential sources of negative energy, but even in these cases, there's no guarantee stable wormholes can occur.

Crossing Cosmic Strings to Allow Time Travel

Cosmic strings are theoretical objects that predate string theory, but in recent years there's been some speculation that they may actually be enlarged strings left over from the big bang, or possibly the result of branes colliding. There has also been speculation that they can be used to create a time machine.

Regardless of their origin, if cosmic strings exist, they should have an immense amount of gravitational pull, and this means that they can cause frame dragging. In 1991, J. Richard Gott (who, with William Hiscock, solved Einstein's field equations for cosmic strings in 1985) realized that two cosmic strings could actually allow time travel.

The way this works is that two cosmic strings cross paths with each other in a certain way, moving at very high speeds. A spaceship traveling along the curves could take a very precise path (several of which were worked out by Curt Cutler in the months after Gott's publication) and arrive at its starting position, in both space and time, allowing for travel in time. Like other time machines, the spaceship couldn't travel further back than when the cosmic strings originally got in position to allow the travel — in essence, the time travel is limited to when the cosmic string time machine was activated.

Gott's was the second time machine (following Kip Thorne's) to have been published in a major journal in the early 1990s, and it sparked a wave of work in the area. In May 1991, Gott was featured in *Time* magazine. In the summer of 1992, physicists held a conference on time travel at the Aspen Center for Physics (the same place where, nearly a decade earlier, John Schwartz and Michael Green had determined that string theory could be consistent).

When Gott proposed this model, cosmic strings were believed to have nothing to do with string theory. In recent years, physicists have grown to believe that cosmic strings, if they exist, may actually be very closely related to string theory.

A Two-Timing Science: String Theory Makes More Time Dimensions Possible

Because relativity showed time as one dimension of space-time and string theory predicts extra space dimensions, a natural question would be whether string theory also predicts (or at least allows for) extra time dimensions. According to physicist Itzhak Bars, this may actually be the case, in a field he calls *two-time physics*. Though still a marginal approach to string theory, understanding this potential extra dimension of time could lead to amazing insights into the nature of time.

Adding a new time dimension

With one time dimension, you have the arrow of time, but with two time dimensions, things become less clear. Given two points along a single time dimension, there's only one path between them. With two time dimensions, two points can potentially be connected by a number of different paths, some of which could loop back on themselves, creating a route into the past.

Most physicists have never looked into this possibility, for the simple fact that (in addition to making no logical sense) it wreaks havoc with the mathematical equations. Time dimensions have a negative sign, and if you incorporate even more of them you can end up with negative probabilities of something happening, which is physically meaningless.

However, Itzhak Bars of the University of Southern California in Los Angeles discovered in 1995 that M-theory allowed for the addition of an extra dimension — as long as that extra dimension was timelike.

To get this to make any sense, he had to apply another type of gauge symmetry, which placed a constraint on the way objects could move. As he explored the equations, he realized that this gauge symmetry only worked if there were two extra dimensions — one extra time dimension and one extra space dimension. Two-time relativity has four space dimensions and two time dimensions, for a total of six dimensions. Two-time M-theory, on the other hand, ends up with 13 total dimensions — 11 space dimensions and two time dimensions.

The gauge symmetry that Bars introduced provided exactly the constraint he needed to eliminate time travel and negative probabilities from his theory. With his gauge symmetry in place, the world with six (or 13) dimensions should behave exactly like the world with four (or 11) dimensions.

Reflecting two-time onto a one-time universe

In a 2006 paper, Bars showed that the Standard Model is a shadow of his 6-dimensional theory. Just like a 2-dimensional shadow of a 3-dimensional object can vary depending on where the light source is placed, the 4-dimensional physical properties ("shadows") can be caused by the behavior of the 6-dimensional objects. The objects in the extra dimensions of Bars's two-time physics theory can have multiple shadows in the 4-dimensional universe (like ours), each of which corresponds to different phenomena. Different physical phenomena in our universe can result from the same fundamental 6-dimensional objects, manifesting in different ways.

To see how this works, consider a particle moving through empty space in six dimensions, with absolutely no forces affecting it. According to Bars's calculations, such activity in six dimensions relates to at least two shadows (two physical representations of this 6-dimensional reality) in the 4-dimensional world:

> ✔ An electron orbiting an atom
>
> ✔ A particle in an expanding universe

Bars believes that two-time physics can explain a puzzle in the Standard Model. Some parameters describing quantum chromodynamics (QCD) have been measured to be quite small, meaning that certain types of interactions are favored over others, but nobody knows why this is. Physicists have come up with a possible fix for this, but it involves predicting a new theoretical particle called an *axion,* which has never been observed.

According to Bars's predictions, two-time physics presents a 4-dimensional world in which QCD interactions are not at all lopsided, so the axion isn't needed. Unfortunately, the lack of discovery of an axion isn't really enough to be counted as experimental proof of two-time physics.

For that, Bars has applied two-time physics to supersymmetry. In this case, the superpartners predicted have slightly different properties than the superpartners predicted by other theories. If superpartners are observed at the Large Hadron Collider with the properties Bars suggests, this would be considered intriguing experimental evidence in favor of his claims.

Does two-time physics have any real applications?

Most physicists believe that these extra-dimensional results from Bars are just mathematical artifacts. Several theorists, including Stephen Hawking, have used the idea of imaginary time dimensions (an imaginary quantity in mathematics is the square root of a negative number), but rarely is this believed to have a real physical existence. To most physicists, they're mathematical tools that simplify the equations.

However, history has shown that "mathematical artifacts" can frequently have a real existence. Bars himself seems to believe that they have as much physical reality as the four dimensions that we know exist, although we'll never experience these extra dimensions as directly.

Though two-time physics doesn't directly imply any time travel, if it's true, it means that time is inherently more complex than physicists have previously believed. Unraveling the mystery of two-time physics could well introduce new ways that time travel might manifest in our universe.

Sending Messages through Time

The original string theory, bosonic string theory, contained a massless particle called the tachyon, which travels faster than the speed of light. In Chapter 10, I explain how these particles are usually a sign that a theory has an inherent flaw — but what if they actually existed? Would they allow a means of time travel?

The short answer is that no one knows. The presence of tachyons in a theory means that things begin to go haywire, which is why they're considered by physicists to be a sign of fundamental instabilities in the theory. (These instabilities in string theory were fixed by including supersymmetry, creating superstring theory — see Chapter 10.)

However, just because tachyons mess up the mathematics that physicists use doesn't necessarily mean that they don't exist. It may be possible that physicists just haven't developed the proper mathematical tools to address them in a way that makes sense.

If tachyons do exist, then in theory it would be possible to send messages that travel faster than the speed of light. These particles could actually travel backward in time and, in principle, be detected.

To avoid this problem (because, remember, time travel can destroy all of physics!), the physicist Gerald Feinberg presented the *Feinberg reinterpretation principle* in 1967, which says that a tachyon traveling back in time can be reinterpreted, under quantum field theory, as a tachyon moving forward in time. In other words, detecting tachyons is the same as emitting tachyons. There's just no way to tell the difference, which would make sending and receiving messages fairly challenging.

Time Travel Paradoxes

Time travel inherently creates a number of logical inconsistencies, called *paradoxes*. These problems have created some of the finest science fiction tales and films (see the nearby sidebar), and have troubled philosophers and scientists since they were first posed. Whether these inconsistencies mean that time travel is physically impossible remains to be seen, although they're among the reasons why most physicists tend to believe that time travel is impossible.

The twin paradox

The *twin paradox* is one of the classic examples of Einstein's theory of relativity in action and dates back nearly as far as the theory itself. It is a thought experiment that exhibits the strange results of time dilation. (Technically, the twin paradox isn't a paradox so much as a problem of inconsistent measurements, but the name has stuck.)

Imagine our pair of twins, Maggie and Emily, once again. At age 20, Emily chooses to become an astronaut, getting recruited onto the first interstellar mission. Her ship is heading to a star that is 10 light-years away, but the ship will be traveling at nearly the speed of light, so time dilation will be in effect.

The ship is truly a wonder, and thanks to time dilation, the entire trip takes only a couple of months from Emily's standpoint. She explores the distant region for eight months, collecting much fascinating data. She then returns, which also takes a couple of months. The entire trip takes Emily one year.

Maggie, on the other hand, stays on Earth. Because Emily was traveling to a star 10 light-years away at nearly the speed of light, Emily arrives at the star when Maggie is about 30 years old. She starts her return trip eight months later, and that leg of the trip also takes 10 years.

The twins have a tearful reunion, where the twin paradox suddenly becomes clear to each of them as the 41-year-old Maggie embraces her twin sister Emily, who appears to be 21 years old. Here is the "paradox:"

> What is Emily's actual age?

After all, Emily was born 41 years ago, the same time Maggie was. There is a logical sense in which Emily is 41 years old. On the other hand, by her "biological clock" only 21 years have passed, and she certainly doesn't look 41.

There isn't a single solution to the twin paradox, because the flow of time depends on how you choose to measure it. Time is, if you'll excuse the phrase, relative.

No doubt if space travel ever becomes feasible, conventions of measuring time will need to be made. For example, if the legal drinking age is 21, can an 18-year-old who has spent four years traveling at near lightspeed buy alcohol legally?

The grandfather paradox

Another paradox is called the *grandfather paradox* and comes up in cases where you can time travel into the past. If you travel into the past, it should be possible to alter the past. The grandfather paradox asks:

> What happens if you change the past in a way that results in you being unable to go into the past in the first place?

Consider the classic example (which gives the paradox its name):

1. You travel into the past and accidentally cause the death of your own grandfather.

2. You cease to exist, and therefore do not travel into the past.

3. You do not cause the death of your grandfather.

4. You now exist, so continue back to step 1.

There are two logical resolutions to the paradox. (See the sidebar "The science fiction of time" for examples of both of these resolutions, in the form of *Back to the Future* and *Somewhere in Time.*)

The first is based off of the many world interpretation (MWI) of quantum physics. In this view, many possible timelines exist and we exist in one of them. If you travel back in time and alter time, then you will continue forward in a different timeline than the one you initially began in.

The second possible resolution to the grandfather paradox is that it's actually impossible to alter the past. The past is set in stone, and if you travel to the past, you'll find that you're unable to change the events that took place, no matter how hard you try.

Unfortunately, the second possibility creates some philosophical problems with free will, because if you are already part of the past, then that means that your own future is set — you will definitely travel into the past at some point. The past and the future both become set in stone.

Of course, no one knows which resolution is correct, and if Stephen Hawking's chronology protection conjecture is true, it's very likely that the situation never will arise. Still, it's fun to speculate on.

Where are the time travelers?

One of the most practical paradoxes brought up regarding time travel is the current lack of any time travelers. If time travel into the past were possible, then it would seem like people from the future would be showing up in our present.

The science fiction of time

Talking about time travel without mentioning science fiction would leave an elephant in the chapter, so to speak. Here are some key science fiction novels and films related to the time travel concepts discussed in this chapter, although the list is by no means complete. Spoiler alert: Some plot details are revealed in the descriptions below.

Novels:

✔ *The Time Machine*, **by H. G. Wells (1895):** The first story with a man-made device to travel in time, where the travel was under the control of the traveler (as opposed to stories that preceded it like *Rip van Winkle*, *A Connecticut Yankee in King Arthur's Court*, or *A Christmas Carol*, where the time traveler had no control).

✔ *Tau Zero*, **by Poul Anderson (1967):** A spaceship is trapped accelerating closer and closer to the speed of light, unable to decelerate. The novel explores the effects of time dilation and the possible end of the universe.

✔ *Gateway*, **by Frederick Pohl (1977):** The sole survivor of a space accident has to come to terms with intense survivor's guilt for the crew he left behind. The plot's powerful climax (which I may now be ruining by telling you) relates to the idea that as you fall into a black hole, time slows down.

Films:

✔ *Somewhere in Time* **(1980):** Richard Collier (Christopher Reeve) is a playwright who travels to 1912 from 1980. The film takes the stance that the past has already happened and Collier was already part of the events of the past (or he's hallucinating, in which case this film has nothing to do with time travel and is far less interesting). For example, "before" he ever time travels, he finds his own signature in a hotel guestbook from 1912. Based on a novel by Richard Matheson.

✔ *Back to the Future* **(1985):** Marty McFly (Michael J. Fox) travels from 1985 to 1955 and interferes with his parents' first meeting. The film explores the concept of time paradoxes and potential multiple timelines. There were two sequels, but the original film was by far the best.

✔ *Frequency* **(2000):** New York detective John Sullivan (James Caviezel) begins communicating with his father (Dennis Quaid) 30 years in the past over a ham radio, which is bouncing signals off of strange sunspot activity. In this film, no material objects travel in time — only information in the form of radio waves. String theorist and author Brian Greene served as physics consultant and had a cameo in the film.

Not only do science fiction authors learn from scientists in developing their time travel systems, but inspiration can flow the other way. Dr. Ronald Mallet, who is trying to build a time machine, was motivated throughout his life by science fiction accounts of time travel. Kip Thorne has developed his theories of time travel out of helping friends work out the details of their science fiction novels. His first work on time travel was based on work performed to help Carl Sagan develop a realistic wormhole for his novel *Contact* in the 1980s, and he later gained insights from the science fiction author Robert Forward.

The solution to this question in the *Star Trek* universe is the "temporal prime directive," which basically makes the argument that time travelers are forbidden from interfering in the past. In this way, any time travelers among us would have to stay hidden.

A more scientific solution is the idea that time travel is only allowed after a time travel device has been constructed. When the device is active, you could use it to travel in time, but you obviously could never go to a time period before the device was created. In fact, every time machine that scientists have found that *could* exist in our universe has this very feature — you can never go back to before the invention (and activation) of the time machine.

Part V

What the Other Guys Say: Criticisms and Alternatives

The 5th Wave   By Rich Tennant

"I just can't help believing that string theory is our best means of understanding this crazy universe we live in."

In this part . . .

Not everyone embraces string theory as the theory that will answer the fundamental questions of physics. In fact, in recent years, even some string theorists have begun backing off of that claim.

In this part, I explain some of the major criticisms of string theory in recent years. I then explore the major alternative quantum gravity theory, loop quantum gravity, and other directions of research that may provide insights, regardless of whether string theory ultimately fails or succeeds. If string theory does fail, or even if it succeeds but not as a "theory of everything," these alternative approaches may prove useful in filling the gaps. Some of these research efforts may provide clues that could help with the progress of string theory.

Chapter 17

Taking a Closer Look at the String Theory Controversy

. . . .

In This Chapter

▶ Considering what string theory does and doesn't explain

▶ Realizing that string theory may never explain our universe

▶ Should string theorists control physics departments and research funding?

. . . .

*A*lthough many physicists believe that string theory holds the promise as the most likely theory of quantum gravity, there's a growing skepticism among some that string theory hasn't achieved the goals it set out for. The major thrust of the criticism is that, whatever useful benefits there are to studying string theory, it's not actually a fundamental theory of reality, but only a useful approximation.

String theorists acknowledge some of these criticisms as valid and dismiss others as premature or even completely contrived. Whether or not the critics are right, they've been a part of string theory since the very first days and are likely to be around as long as the theory persists. Lately, the criticism has risen to such furor that it's being called "the string wars" across many science blogs and magazines.

In this chapter, I discuss some of the major criticisms of string theory. I begin with a brief recap of the history of string theory, from the eye of the skeptic, who focuses on the failures instead of the successes. After that, I look into whether string theory has any ability to actually provide any solid predictions about the universe. Next, you see how string theory critics object to the extreme amount of control that string theorists hold over academic institutions and research plans. I then consider whether string theory possibly describes our own reality. And, finally, I explain some of the major string theory responses to these criticisms.

The String Wars: Outlining the Arguments

As long as it's been around, string theory has contended with criticisms. Some of string theory's critics are among the most respected members of the physics community, including Nobel laureates such as Sheldon Glashow and the late Richard Feynman, both of whom were critical as far back as the first superstring revolution in the mid-1980s. Still, string theory has steadily grown in popularity for decades.

Recently, the rise in criticisms against string theory has spilled into the popular media, making the front pages of science magazines and even large articles in more mainstream publications. The debate rages across radio waves, the Internet, academic conferences, the blogosphere, and anywhere else that debates are allowed to rage.

Though the debate sounds passionate, none of the critics are really advocating that physicists completely abandon string theory. Instead, they tend to view string theory as an *effective theory* (a useful approximation) rather than a truly *fundamental theory,* which describes the most basic level of reality itself. They are critical of string theorists' attempts to continue to promote the theory as a fundamental theory of reality.

Here are some of the most significant criticisms levied against string theory (or the string theorists who practice it):

✔ String theory is unable to make any useful prediction about how the physical world behaves, so it can't be falsified or verified.

✔ String theory is so vaguely defined and lacking in basic physical principles that any idea can be incorporated into it.

✔ String theorists put too much weight on the opinions of leaders and authorities within their own ranks, as opposed to seeking experimental verification.

✔ String theorists present their work in ways that falsely demonstrate that they've achieved more success than they actually have. (This isn't necessarily an accusation of lying, but may be a fundamental flaw in how success is measured by string theorists and the scientific community at large.)

✔ String theory gets more funding and academic support than other theoretical approaches (in large part because of the aforementioned reported progress).

✔ String theory doesn't describe our universe, but contradicts known facts of physical reality in a number of ways, requiring elaborate hypothetical constructions that have never been successfully demonstrated.

Behind many of these criticisms is the assumption that string theory, which has been around for 30 years, should be a bit more fully developed than it actually is. None of the critics are arguing to abandon the study of string theory; they just want alternative theories to be pursued with greater intensity, because of the belief that string theory is falling short of the mark.

To explore the validity of these claims and determine whether string theory is in fact unraveling, it's necessary to lay out the frame of the debate by looking at where string theory has been and where it is today.

Thirty years and counting: Framing the debate from the skeptic's point of view

Even now, with criticism on the rise, it doesn't appear that the study of string theory has dropped. To understand why physicists continue to study string theory, and why other physicists believe it isn't delivering as promised, let me briefly recount the general trends in the history of string theory, focusing this time on its shortcomings. (This material is presented in significantly greater detail in Chapters 10 and 11.)

String theory started in 1968 as a theory (called the dual resonance model) to predict the interactions of hadrons (protons and neutrons), but failed at that. Instead of this model, quantum chromodynamics, which said that hadrons were composed of quarks held together by gluons, proved to be the correct model.

Analysis of the early version of string theory showed that it could be viewed as very tiny strings vibrating. In fact, this bosonic string theory had several flaws: fermions couldn't exist and the theory contained 25 space dimensions, tachyons, and too many massless particles.

These problems were "fixed" with the addition of supersymmetry, which transformed bosonic string theory into superstring theory. Superstring theory still contained nine space dimensions, though, so most physicists still believed it had no physical reality.

This new version of string theory was shown to contain a massless, spin-2 particle that could be the graviton. Now, instead of a theory of hadron interactions, string theory was a theory of quantum gravity. But most physicists were exploring other theories of quantum gravity, and string theory languished throughout the 1970s.

The first superstring revolution took place in the mid-1980s, when physicists showed ways to construct string theory that made all the anomalies go away. In other words, string theory was shown to be consistent. In addition, physicists found ways to compactify the extra six space dimensions by curling them up into complex shapes that were so tiny they would never be observed.

The rise in work on string theory had great results. In fact, the results were too good, because physicists discovered five distinct variations of string theory, each of which predicted different phenomena in the universe and none of which precisely matched our own.

In 1995, Edward Witten proposed that the five versions of string theory were different low-energy approximations of a single theory, called M-theory. This new theory contained ten space dimensions and strange objects called branes, which had more dimensions than strings.

A major success of string theory was that it was used to construct a description for black holes, which calculated the entropy correctly, according to the Hawking-Bekenstein predictions for black hole thermodynamics. This description applied only to specific types of simplified black holes, although there was some indication that the work might extend to more general black holes.

A problem for string theory arose in 1998, when astrophysicists showed that the universe was expanding. In other words, the cosmological constant of the universe is positive, but all work in string theory had assumed a negative cosmological constant. (The positive cosmological constant is commonly referred to as dark energy.)

In 2003, a method was found to construct string theory in a universe that had dark energy, but there was a major problem with it: A vast number of distinct string theories were possible. Some estimates have been as much as 10^{500} distinct ways to formulate the theory, which is so absurdly large that it can be treated as if it were basically infinity.

As a response to these findings, physicist Leonard Susskind proposed the application of the anthropic principle as a means of explaining why our universe had the properties it did, given the incredibly large number of possible configurations, which Susskind called the landscape.

This brings us to the current status of string theory, in very broad strokes. You can probably see some chinks in string theory's armor, where the criticisms seem to resonate particularly strongly.

A rise of criticisms

After evidence of dark energy was discovered in 1998 and the 2003 work increased the number of known solutions, there seemed to be some growth in criticisms. The attempts to make the theory fit physical reality were growing a bit more strained, in the eyes of some, and a discontent that had always existed under the surface began to seep out of the back rooms at physics conferences and onto the front pages of major science magazines.

While innovative new variants — such as the Randall-Sundrum models and the incorporation of a positive cosmological constant — were rightly recognized as brilliant, some people believed that physicists had to come up with contrived explanations to keep the theory viable.

The growth in criticism became glaringly obvious to the general public in 2006 with the publication of two books criticizing — or outright attacking — string theory. The books were Lee Smolin's *The Trouble with Physics: The Rise of String Theory, the Fall of a Science, and What Comes Next* and Peter Woit's *Not Even Wrong: The Failure of String Theory and the Search for Unity in Physical Law.* These books, along with the media fervor that accompanies any potential clash of ideas, has put string theory on the public relations defensive even while many (possibly most) string theorists dismiss the Smolin and Woit claims as failed attempts to discredit string theory for their own aggrandizement.

The truth is likely somewhere in between. The criticisms have a bit more merit than string theorists would give them, but are not quite as destructive as Woit, at least, would tend to have readers believe. (Smolin is a bit more sympathetic toward string theory, despite his book's subtitle.) None of the critics propose abandoning string theory entirely; they merely would like to see more scientists pursuing other areas of inquiry, such as those described in Chapters 18 and 19.

Is String Theory Scientific?

The first two criticisms cut to the core of whether string theory is successful as a scientific theory. Not just any idea, not even one that's expressed in mathematical terms, is scientific. In the past, to be scientific, a theory had to describe something that is happening in our own universe. To go too far from this boundary enters the realm of speculation. Criticisms of string theory as unscientific tend to fall in two (seemingly contradictory) categories:

- ✔ String theory explains nothing.
- ✔ String theory explains too much.

Argument No. 1: String theory explains nothing

The first attack on string theory is that, after about 30 years of investigation, it still makes no clear predictions. (Physicists would say it has no *predictive power.*) The theory makes no unique prediction that, if true, supports the theory and, if false, refutes the theory.

According to philosopher Sir Karl Popper, the trait of "falsifiability" is the defining trait of science. If a theory is not falsifiable — if there is no way to make a prediction that gets a false result — then the theory is not scientific.

If you subscribe to Popper's view (and many scientists don't), then string theory is certainly not scientific — at least not yet. The question is whether string theory is fundamentally unable to make a clear, falsifiable prediction or whether it merely hasn't done so yet, but will at some point in the future.

It's possible that string theorists will make a distinct prediction at some point. Part of the criticism, though, is that string theorists are really not concerned with making a prediction. Some string theorists don't even seem to consider the lack of a currently testable prediction to be a shortcoming, so long as string theory remains consistent with the known evidence.

This is what motivates the major critics of string theory, from Feynman in the 1980s to Smolin and Woit today, to complain that string theory has no contact with experiment and is fundamentally warping what it means to investigate something scientifically.

Argument No. 2: String theory explains too much

The second attack is based on the same problem, that string theory makes no unique prediction, but the emphasis this time is on the word "unique." There are so many variations of string theory that even if it could be formulated in a way that it would make a prediction, it seems as if each version of string theory would make a slightly different prediction.

This is, in a way, almost worse than making no prediction at all. With no prediction, you can make the argument that more work and refinement needs to be done, new mathematical tools developed, and so on. With a nearly infinite number of predictions, you're stuck with a theory that's completely useless. Again, it has no predictive power, for the simple reason that you can never sort out the sheer volume of results.

Part of this argument relates back to the principle of Occam's razor. According to this principle, there is an *economy* in nature, which means that nature (as described by science) doesn't include things that aren't necessary. String theory includes extra dimensions, new types of particles, and possibly whole extra universes that have never been observed (and possibly *can never* be observed).

New rules to the game: The anthropic principle revisited

The solution for so many predictions, as proposed by physicist Leonard Susskind, is to apply the anthropic principle to focus on the regions of the string theory landscape that allow life to exist. According to Susskind, Earth clearly exists in a universe (or a region of the universe, at least) that allows life to exist, so selecting only theories that allow life to exist seems to be a reasonable strategy.

Taking a theory that doesn't allow life to exist and considering it on equal footing with theories that do allow life to exist, when we know that life *does* exist, defies both scientific reasoning and common sense.

From this stance, the anthropic principle is a way of removing selection bias when looking at different possible string theories. Instead of looking only at the mathematical viability of a theory, as if that were the only criteria, physicists can also select based on the fact that we live here.

However, there's a bit of clever maneuvering within this discussion that shouldn't go unmentioned. It's not just that Susskind has said that we can use the anthropic principle to select which theories are viable in our universe, but he's gone further to indicate that the very fact that all of these versions of string theory exist is a *good* thing. It provides a richness to the theory, making it more robust. (Still others point out that all quantum field theories have lots of potential solutions, so string theory shouldn't be any different. In those cases, both sides of this particular debate are looking at it the wrong way.)

For nearly two decades, many physicists were trying to find a single version of string theory that included basic physical principles that dictated the nature of the universe. The current Standard Model has 18 fundamental

particles, which have to be measured in experiment and placed into the theory by hand. Part of the goal of string theory was to find a theory that, based on pure physical principles and mathematical elegance, would yield a single theory describing all of reality.

Instead, string theorists have found a virtually infinite number of different theories (or, to be more precise, different string theory solutions) and have apparently discovered that no fundamental law describes the universe based on basic physical principles. Selection of the correct parameters for the theory is, once again, left to experiment.

But instead of interpreting this as a failure and indicating that we have no choice but to apply the anthropic principle to provide limitations on which options are available to us, Susskind takes lemons and turns them into lemonade by reframing the entire context of success. Success is no longer finding a single theory, but exploring as much of the landscape as possible.

In their book *Aristotle and an Aardvark go to Washington: Understanding Political Doublespeak Through Philosophy and Jokes,* authors Thomas Cathcart and Daniel Klein refer to this sort of technique as the "Texas sharpshooter fallacy." Imagine the Texas sharpshooter who pulls out his pistol and fires at the wall and then walks up and draws the bull's-eye around the location where the shots landed.

In a (very critical) sense, this is what Susskind has done, by changing the actual definition of success in string theory. He has (according to some) redefined the goal of the enterprise and done so in such a way that the current work is exactly in line with the new goal. If this new approach is valid, yielding a way to correctly describe nature, it's brilliant. If not valid, then it's not brilliant. (For the more favorable interpretation of the anthropic principle, see Chapters 11 and 14.)

A similar moving target can be seen in the discussion of proton decay. Originally, experiments to prove grand unified theories (GUTs) anticipated that these experiments would detect the decay of a few protons every year. No proton decays have been found, however, which has caused theorists to revise their calculations to arrive at a lower decay rate. Except most physicists believe that these attempts are not valid and that these GUT approaches have been disproved. This after-the-fact change in what they're looking for is not a valid approach to science — unless the decays are discovered at the new rate, of course (at which point the theoretical modification becomes a brilliant insight).

None of this is to imply that Susskind is being dishonest or manipulative in presenting the anthropic principle as an option that he believes in. He has very genuinely been led to this belief because of the growing number of mathematically viable string theory solutions, which leave him with no choice (except for abandoning string theory, which I get to in a bit).

After you accept that string theory dictates a large number of possible solutions, and you realize that modern theories of eternal inflation dictate that many of these solutions may well be realized in some reality, there's very little choice, in Susskind's view, other than to accept the anthropic principle. And there's every indication that he went through some serious soul searching before deciding to preach the anthropic message.

Interpreting the string theory landscape

No longer is string theory looking for a single theory, but it's now trying to pare down the vast options in the landscape to find the one, or the handful, that may be consistent with our universe. The anthropic principle can be used as one of the major selection criteria to distinguish theories that clearly don't apply to our universe.

The question that remains is whether string theorists (or any physicist) should be happy about this situation.

Certainly, some are not. David Gross is not. Edward Witten seems at best lukewarm about the prospect. Susskind and Joe Polchinski, however, seem to have had a full conversion. They have not only resigned themselves to accepting the circumstances, but have embraced it, despite the fact that a few years ago both were opposed to any application of the anthropic principle in science.

The anthropic principle seems unavoidable if there exists a vast multiverse, where many different regions of the string theory landscape are realized in the form of parallel universes. Some universes will exist where life is allowed, and we're one of them — get used to it.

Some string theorists who haven't accepted the anthropic arguments are hopeful that the theory's mathematical and physical features can rule out large portions of the landscape. String theorists are still divided over exactly what conclusions the theory allows and whether there might be some way to sort them out without applying the anthropic principle. More work must be done before anyone knows for sure.

Turning a Critical Eye to String Theorists

One of the major criticisms of string theory has to do not with the theory so much as with theorists. The argument is that they are forming something of a "cult" of string theorists, who have bonded together to promote string theory above all alternatives.

This criticism, which is at the heart of Smolin's *The Trouble with Physics,* is not so much a criticism of string theory as a fundamental criticism of the way academic resources are allocated. One criticism of Smolin's book has been that he is in part demanding more funding for the research projects that he and his friends are working on, which he feels are undersupported. (Many of these alternative fields are covered in Chapters 18 and 19.)

Hundreds of physicists just can't be wrong

String theory is the most popular approach to a theory of quantum gravity, but that very phrase — most popular — is exactly the problem in the eyes of some. In physics, who cares (or who *should* care) how popular a theory is?

In fact, some critics believe that string theory is little more than a cult of personality. The practitioners of this arcane art have long ago foregone the regular practice of science, and now bask in the glory of seer-like authority figures like Edward Witten, Leonard Susskind, and Joe Polchinski, whose words can no more be wrong than the sun can stop shining.

This is, of course, an exaggeration of the criticism, but in some cases, not by much. String theorists have spent more than two decades building a community of physicists who firmly believe that they are performing the most important science on the planet, even while achieving not a single bit of evidence to definitively support their version of science as the right one, and the folks at the top of that community carry a lot of weight. (For a look at this behavior in nonphysics contexts, see the nearby sidebar "Appeal to authority.")

John Moffat has joined Smolin and Woit in lamenting the "lost generation" of brilliant physicists who have spent their time on string theory, to no avail. He points out that the sheer volume of physicists publishing papers on string theory, and in turn citing other string theorists, skews the indexes about which papers and scientists are truly the most important.

For example, there is a rumor that Edward Witten has the highest h-index of any living scientist. (The *h-index* is a measure of how often papers are cited.) If you look at it from Moffat's point of view, this is not necessarily a result of Witten being the most important physicist of his generation, but rather a result of Witten writing papers that are fundamental to string theory, and, in turn, are cited by the vast majority of people writing papers on string theory, which is a lot of papers.

Now the problem with this approach when it comes to Witten specifically is that it's very possible that he *is* the most important physicist of his generation. Certainly his Fields Medal attests to his position as one of the most mathematically gifted. But if he is an important physicist who has helped lead a generation of physicists down a road that ends in string theory as a failed theory of quantum gravity, then that would indeed make for a "lost generation" and a tragic waste of Witten's brilliance.

Holding the keys to the academic kingdom

The theoretical physics and particle physics communities in many of the major physics departments, especially in the United States, lean heavily toward string theory as the preferred approach to a quantum gravity theory. In fact, the growing need for diverse approaches (such as those from Chapters 18 and 19) is maintained even by some string theorists, who realize the importance of including conflicting viewpoints.

Appeal to authority

Although it may seem odd to many people that scientists could be swayed by figures of authority, this is a fundamental part of human nature. The "appeal to authority" was cited by Aristotle, the father of rhetoric (the science of debate). It has been given the Latin name *argumentum ad verecundiam,* and evidence from psychology has born out that it works. People are inclined to believe an authority figure, sometimes even over common sense.

Marketers know that one of the most persuasive ways to sell something is to get a testimonial. This is why speakers are introduced by someone else, for example. If another person gets up and lists the speaker's accomplishments, it means a lot more to the listeners than if the speaker stands up, introduces herself, and lists off her own accomplishments. This is the case even when the introducer knows nothing about the person except what he reads off of a card or teleprompter.

When the person who is providing the testimonial is perceived as an authority figure, it's even more potent. This is why some books have quotes from authorities on them and why politicians seek celebrity endorsements. I'm sure some people voted for Barack Obama in 2008 because Oprah Winfrey, an authority figure if ever there was one, endorsed him publicly.

In the case of string theory, of course, the authority figures aren't just popular, they are experts in physics, and string theory in particular, so listening to their opinion on string theory is a bit more reasonable than listening to a single popular actor, musician, athlete, or clergyman on whom to vote into the presidency. Ultimately, in science (as in the rest of life) people should use their own logic to evaluate the arguments put forward by the experts. Fortunately, scientists are trained to use their logic more intently than most of society.

In a debate between Brian Greene and Lee Smolin on National Public Radio, Greene acknowledged the need to work on areas other than string theory, pointing out that some of his own graduate students are working on other approaches to solving problems of quantum gravity.

Lisa Randall — whose own work has often been influenced by string theory — describes how, during the first superstring revolution, Harvard physicists remained more closely tied to the particle physics tradition, and to experimental results, while Princeton researchers devoted themselves largely to the purely theoretical enterprise of string theory. In the end, every particle theorist at Princeton worked on string theory, which she identifies as a mistake — and one that continues to this day.

These stances indicate that if a "string theory cult" does exist, then Brian Greene and Lisa Randall have apparently not been inducted into it. Still, the fact is that theoretical physics departments at several major universities are now dominated by string theory supporters, and some feel that other approaches are inherently marginalized by that.

This criticism is one of the fairest, I think, because science, like any other field of endeavor, *needs* criticism. Psychologists have shown that the phenomenon of "groupthink" takes hold in situations where the only people who are allowed a seat at the table are those who think alike. If you want to have a robust intellectual exchange — something that's at the heart of physics and other sciences — it's important that you include people who will challenge your viewpoints and not just agree with them.

Some criticisms of Smolin's book have indicated that he wants some sort of handout for himself and his buddies who aren't able to cut it in the normal grant application process. (In the other direction, Smolin and Woit have implied that similar economic interests are at the heart of the support for string theory.)

But if the institutes that determine how funding is allocated are dominated by people who believe that string theory is the only viable theory, then these alternate approaches won't get funded. Add to that the citation issues described earlier in this chapter, which possibly make string theory look more successful than it actually is, and there's room for valid criticism of how funding is allocated in physics.

Still, hope for these alternatives isn't lost. As popular as string theory is, I believe it's likely that most theoretical physicists want to find answers more than they want to be proved right. Physicists will gravitate (so to speak) toward the theories that provide them the best opportunity to discover a fundamental truth about the universe.

So long as these non-string theorists continue doing solid work in these other areas, then they have the hope of drawing recruits from the younger generation. Eventually, if string theorists don't find some way to make string theory succeed, it will lose its dominant position.

Does String Theory Describe Our Universe?

Now comes the real science question related to string theory: Does it describe our universe? The short answer is that no, it does not. It can be written in such a way to describe some idealized worlds that bear similarities to our world, but it can't yet describe our world.

Unfortunately, you have to know a lot about string theory to realize that. String theorists are rarely upfront about how far their theory is from describing our reality (when talking to public audiences, at least). It tends to be a disclaimer, woven into the details of their presentations or thrown in just near the end. In fact, you could read many of the books out there on string theory and, after turning the last page, you wouldn't have ever been told explicitly that it doesn't describe our universe.

Congratulations on not choosing one of those books.

Making sense of extra dimensions

The world described by string theory has at least 6 more space dimensions than the 3 we know, for a total of 9 space dimensions. In M-theory, there are at least 10 space dimensions, and in the two-time M-theory, there are 11 space dimensions (with 2 time dimensions tacked on).

The problem is that physicists don't know where these extra dimensions are. In fact, the main reason for believing that they exist is that the equations of string theory demand them. These extra dimensions have been compactified (in some models) in ways that their particular geometry generates certain features of our universe.

There are two major ways of dealing with the extra dimensions:

- ✔ The extra dimensions are compactified, probably at about the Planck scale (although some models allow for them to be larger).
- ✔ Our universe is "stuck" on a three space dimensional brane (brane world scenarios).

There is another alternative: The extra dimensions may not exist. (This would be the approach suggested by applying Occam's razor.) Various physicists have developed approaches to string theory without extra dimensions, as discussed in Chapter 13, so abandoning the idea of extra dimensions doesn't even require an abandonment of string theory!

Space-time should be fluid

One of the hallmarks of modern physics is general relativity. The clash between general relativity and quantum physics is part of the motivation for looking for a string theory, but some critics believe that string theory is designed in such a way that it doesn't faithfully maintain the principles of general relativity.

Which principles of general relativity aren't maintained in string theory? Specifically, the idea that space-time is a dynamic entity that responds to the presence of matter around it. In other words, space-time is flexible. In physics terminology, general relativity is a *background-independent theory,* because the background (space-time) is incorporated into the theory. A *background-dependent theory* is one where objects in the theory are sort of "plugged in" to a space-time framework.

Right now, string theory is a background-dependent framework. Space-time is rigid, instead of flexible. If you are given a certain configuration of space-time, you can discuss how a given version of string theory would behave in that system.

The question is whether string theory, which right now can only be formulated in fixed space-time environments, can really accommodate a fundamentally dynamic space-time framework. How can you turn the rigid space-time of string theory into the flexible space-time of general relativity? The pessimist replies "You can't" and works on loop quantum gravity (see Chapter 18).

The optimist, however, believes that string theory still has hope. Even with a rigid background of space-time, it's possible to get general relativity as a limiting case of string theory. This isn't quite as good as getting a flexible space-time, but it means that string theory certainly doesn't exclude general relativity. Instead of getting the full high-definition version of space-time, though, you're left with something more like a flipbook, which treats each image as static but, overall, provides the impression of smooth motion.

String theory is a work-in-progress, and it's still hoped that physical and mathematical principles might be developed that will allow for the expression of a fully dynamic background in string theory. String theorists are forced to talk about the theory in a rigid space-time (background dependent) only because they haven't yet found a mathematical language that will let them talk about it in a flexible space-time (background-independent). Some believe that Maldacena's AdS/CFT correspondence may provide a means of incorporating this background-independent language. It's also possible that the principles that allow this new language will come from an unexpected direction, such as the work described in Chapters 18 and 19.

Or, of course, such principles may not exist at all, and the skeptic's inclination to criticize string theory may therefore be justified.

How finite is string theory?

One criticism that has arisen largely since Smolin's *The Trouble with Physics* is the notion that string theory isn't necessarily a finite theory. Remember that this is one of the key features in support of string theory: It removes the infinities that arise when you try to apply quantum physics directly to problems.

As Smolin describes things, this belief in string theory finiteness is largely based on a 1992 proof performed by Stanley Mandelstam, in which Mandelstam only proved that the first term of string theory (remember that string theory is an equation made up of an infinite series of mathematical terms) was finite. It has since been proved for the second term, as well.

Still, even if every individual term is finite, string theory currently is written in a form (like quantum field theory) that has an infinite number of terms. Even if each term is finite, it's possible that the sum of all of the terms will yield an infinite result. Because infinities are never witnessed in our universe, this would mean that string theory doesn't describe our universe.

The fact that string theory finiteness hasn't been proved isn't a flaw in string theory. The fact that most string theorists *thought* that it had been proved finite when it wasn't is the flaw — not necessarily a flaw in string theory itself, but a flaw in the very way these scientists are practicing their science. The bigger issue at stake in this particular criticism is one of precision and intellectual honesty.

A String Theory Rebuttal

In the light of all of these criticisms, many of which have some measure of validity or logic to them, you may be wondering how anyone could continue working on string theory. How could some of the most brilliant physicists in the world devote their careers to exploring a field that is apparently a house of cards?

The short answer, stated in various forms by many string theorists over the years, is that they find it hard to believe that such a beautiful theory would not apply to the universe. String theory describes all of the behavior of the universe from certain fundamental principles as the vibrations of 1-dimensional strings and compactification of extra dimensional geometries, and can be used in some simplified versions to solve problems that have meaning to physicists, such as black hole entropy.

Most string theorists are able to dismiss the idea that string theory should be further along than it is. String theory does, after all, explore energies and sizes beyond our current technology to test. And, even in cases where experiment can guide theory, there are cases where 30 years was not enough time.

The theory of light took much longer than 30 years to develop. In the late 1600s, Newton described light as tiny particles. In the 1800s, experiments revealed that it traveled as waves. In 1905, Einstein proposed the quantum principles that led to wave-particle duality, which in turn resulted in the theory of quantum electrodynamics in the 1940s. In other words, the rigorous physical examination of light traces a path from Newton through to Feynman that covers about 250 years, filled with many false leads along the way.

And quantum electrodynamics is a quantum field theory, which means it has infinite solutions unless it goes through a process of renormalization. The fact that string theory may also be infinite isn't seen as a big deal, because the existing theory is definitely infinite. (Although, again, one of the motivations of string theory was to *remove* the infinities.)

For that matter, it took more than 1,500 years for heliocentric models of the Earth's motion to be accepted over geocentric models, even though anyone could look up at the sky! It's only because our modern world moves so fast that we feel we need quick and easy answers to something as simple as the fundamental nature of the universe.

As mentioned earlier, neither side has won the debate (or "string wars") yet, but many feel that the very fact that the debate is taking place is, on the whole, good for science. And those who don't — well, they're probably part of a groupthink cult of string theorists.

Chapter 18

Loop Quantum Gravity: String Theory's Biggest Competitor

In This Chapter

▶ Seeing how loop quantum gravity is more focused than string theory

▶ Knowing what predictions loop quantum gravity makes

▶ Evaluating the similarities between loop quantum gravity and string theory

Though string theory is often promoted as the "only consistent theory of quantum gravity" (or something along those lines), some would disagree with this categorization. Foremost among them are the researchers in a field known as *loop quantum gravity* (sometimes abbreviated *LQG*). I discuss other approaches to quantum gravity in Chapter 19.

In this chapter, I introduce you to loop quantum gravity, an alternative theory of quantum gravity. As string theory's major competitor, loop quantum gravity hopes to answer many of the same questions by using a different approach. I start by describing the basic principles of loop quantum gravity and then present some of the major benefits of this approach over string theory. I lay out some of the preliminary predictions of loop quantum gravity, including possible ways to test it. Finally, I consider whether loop quantum gravity has the same fundamental flaws that may bring down string theory.

Taking the Loop: Introducing Another Road to Quantum Gravity

Loop quantum gravity is string theory's biggest competitor. It gets less press than string theory, in part because it has a fundamentally more limited goal: a quantum theory of gravity. Loop quantum gravity performs this feat by trying to quantize space itself — in other words, treat space like it comes in small chunks.

In contrast, string theory starts with methods of particle physics and frequently hopes to not only provide a method of creating a quantum theory of gravity, but also explain all of particle physics, unifying gravity with the other forces at the same time. Oh, and it predicts extra dimensions, which is very cool!

It's no wonder that loop quantum gravity has more trouble getting press.

The great background debate

The key insight of quantum physics is that some quantities in nature come in multiples of discrete values, called *quanta*. This principle has successfully been applied to all of physics, except for gravity. This is the motivation for the search for quantum gravity.

Alternately, the key insight from general relativity is that space-time is a dynamic entity, not a fixed framework. String theory is a *background-dependent theory* (built on a fixed framework; see Chapter 17 for more on this), so it doesn't currently account for the dynamic nature of space-time at the heart of relativity.

According to the LQG researchers, a theory of quantum gravity must be background-independent, a theory that explains space and time instead of being plugged into an already-existing space-time stage. No background-dependent theory can ever yield general relativity as a low-energy approximation.

Loop quantum gravity tries to achieve this goal by looking at the smooth fabric of space-time in general relativity and contemplating the question of whether, like regular fabric, it might be made up of smaller fibers woven together. The connections between these quanta of space-time may yield a background-independent way of looking at gravity in the quantum world.

What is looping anyway?

Loop quantum gravity's key insight is that you can describe space as a *field;* instead of a bunch of points, space is a bunch of lines. The *loop* in loop quantum gravity has to do with the fact that as you view these field lines (which don't have to be straight lines, of course), they can loop around and through each other, creating a *spin network*. By analyzing this network of space bundles, you can supposedly extract results that are equivalent to the known laws of physics.

The foundation of LQG took place in 1986, when Abhay Ashtekar rewrote general relativity as a series of field lines instead of a grid of points. The result turns out not only to be simpler than the earlier approach, but is similar to a gauge theory.

There's one problem, though: Gauge theories are background-dependent theories (they are inserted into a fixed space-time framework), but that won't work, because the field lines themselves represent the geometry of space. You can't plug the theory into space if space is already part of the theory!

In order to proceed, physicists working in this area had to look at quantum field theory in a whole new way so it could be approached in a background-independent setting. Much of this work was performed by Ashtekar, Lee Smolin, Ted Jacobson, and Carlo Rovelli, who can reasonably be considered among the fathers of loop quantum gravity.

As LQG developed, it became clear that the theory represented a network of connected *quantum space bundles,* often called "atoms" of space. The failure of previous attempts to write a quantum theory of gravity was that space-time was treated as continuous, instead of being quantized itself. The evolution of these connections is what provides the dynamic framework of space (although it has yet to be proved that loop quantum gravity actually reduces to the same predictions as those given by relativity).

Each atom of space can be depicted with a point (called a *node*) on a certain type of grid. The grid of all of these nodes, and the connections between them, is called a *spin network.* (Spin networks were originally developed by Oxford physicist Roger Penrose back in the 1970s.) The graph around each node can change locally over time, as shown in Figure 18-1 (which shows the initial state [a] and the new state it changes into [b]). The idea is that the sum total of these changes will end up matching the smooth space-time predictions of relativity on larger scales. (That last bit is the major part that has yet to be proved.)

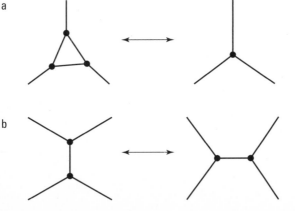

Figure 18-1:
The spin network evolves over time through local changes.

a

b

Now, when you look at these lines and picture them in three dimensions, the lines exist inside of space — but that's the wrong way to think about it. In LQG, the spin network with all of these nodes and grid lines, the entire spin network, is actually space itself. The specific configuration of the spin network is the geometry of space.

The analysis of this network of quantum units of space may result in more than physicists bargained for, because recent studies have indicated that the Standard Model particles may be implicit in the theory. This work has largely been pioneered by Fotini Markopoulou and work by the Australian Sundance O. Bilson-Thompson. In Bilson-Thompson's model, the loops may braid together in ways that could create the particles, as indicated in Figure 18-2. (These results remain entirely theoretical, and it remains to be seen how they work into the larger LQG framework as it develops, or whether they have any physical meaning at all.)

Electron neutrino   Electron anti-neutrino

Positron   Electron

Figure 18-2:
Braids in
the fabric of
space may
account for
the known
particles of
physics.

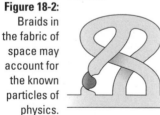

Down quark   Up quark

Making Predictions with Loop Quantum Gravity

Loop quantum gravity makes some definite predictions, which may mean that it could be tested well before string theory can be. As string theory's popularity is being brought into question, the amount of research into LQG may end up growing.

Gravity exists (Duh!)

Oddly enough, because LQG was born out of general relativity, one question has been whether science can get general relativity back out of the theory. In other words, can scientists use loop quantum gravity to actually match Einstein's classical theory of gravity on large scales? The answer is: yes, in some special cases (as does string theory).

For example, work by Carlo Rovelli and his colleagues has shown that LQG contains gravitons, at least in the low-energy version of the theory, and also that two masses placed into the theory will attract each other in accord with Newton's law of gravity. Further theoretical work is needed to get solid correlations between LQG and general relativity.

Black holes contain only so much space

Loop quantum gravity's major success has been in matching the Bekenstein prediction of black hole entropy as well as the Hawking radiation predictions (both described in Chapter 9). As mentioned in Chapters 11 and 14, string theory has been able to make some predictions about special types of black holes, which is also consistent with the Bekenstein-Hawking theories. So, at the very least, if scientists are able to create miniature black holes in the Large Hadron Collider and observe Hawking radiation, then it would certainly not rule out either of the theories.

However, the picture given by LQG is very different from that of classical black holes. Instead of an infinite singularity, the quantum rules say there's only so much space inside of the black hole. Some LQG theorists hope they can predict tiny adjustments to Hawking's theory that, if experimentally proven true, would support LQG above string theory.

One prediction is that instead of a singularity, the matter falling into a black hole begins expanding into another region of space-time, consistent with some earlier predictions by Bryce DeWitt and John Archibald Wheeler. In fact, singularities at the big bang are also eliminated, providing another possible eternal universe model. (For more eternal universe models, see Chapter 14.)

Gamma ray burst radiation travels at different speeds

Many of the experiments from Chapter 12, which could test whether the speed of light varies, would also be consistent with loop quantum gravity. For example, it's possible that gamma ray burst radiation doesn't all travel at the same speed, like classical relativity predicts. As the radiation passes through the spin network of quantized space, the high-energy gamma rays would travel slightly slower than the low-energy gamma rays. Again, these effects would be magnified over the vast distances traveled to possibly be observed by the Fermi telescope.

Finding Favor and Flaw with Loop Quantum Gravity

As with string theory, loop quantum gravity is passionately embraced by some physicists and dismissed by others. The physicists who study it believe that the predictions (described in the preceding section) are far better than those made by string theory. One major argument in support of LQG is that it's seen by its adherents as a finite theory, meaning that the theory itself doesn't inherently admit infinities. These same researchers also tend to dismiss the flaws as being the product of insufficient work (and funding) devoted to the theory. String theorists, in turn, view them as much a victim of "groupthink" as critics view string theorists.

The benefit of a finite theorem

One major benefit of loop quantum gravity is that the theory has been proved finite in a more definitive sense than string theory has. Lee Smolin, one of the key (and certainly most high profile) researchers of LQG, describes in his book *The Trouble with Physics* three distinct ways that the theory is finite (with string theorist objections in parentheses):

✔ The areas and volumes in loop quantum gravity are always in finite, discrete units. (String theorists would say this isn't a particularly meaningful form of finiteness.)

✔ In the Barrett-Crane model of loop quantum gravity, the probabilities for a quantum geometry to evolve into different histories are always finite. (This sounds just like unitarity, which is a property of string theory and all quantum field theories.)

✔ Including gravity in a loop quantum gravity theory that contains matter theory, like the Standard Model, involves no infinite expressions. If gravity is excluded, you have to do some tinkering to avoid them. (String theorists believe this claim is premature and that there are substantial problems with the proposed LQG models that yield this result.)

As I explain in Chapter 17, some questions exist (largely brought up by loop quantum gravity theorists) about whether string theory is actually finite — or, more specifically, over whether it has been rigorously proved finite. From the theoretical side of things, the loop quantum gravity people view this uncertainty as a major victory over string theory. (String theorists would argue that the statements above still don't prove that LQG can't result in an infinite solution when experimental data is put into the theory.)

Spending some time focusing on the flaws

Many of the flaws in loop quantum gravity are the same flaws in string theory. Their predictions generally extend into realms that aren't quite testable yet (although LQG is a bit closer to being able to be experimentally tested than string theory probably is). Also, it's not really clear that loop quantum gravity is any more falsifiable than string theory. For example, the discovery of supersymmetry or extra dimensions won't disprove loop quantum gravity any more than the failure to detect them will disprove string theory. (The only discovery that I think LQG would have a hard time overcoming would be if black holes are observed and Hawking radiation proves to be false, which would be a problem for *any* quantum gravity theory, including string theory.)

The biggest flaw in loop quantum gravity is that it has yet to successfully show that you can take a quantized space and extract a smooth space-time out of it. In fact, the entire method of adding time into the spin network seems somewhat contrived to some critics, although whether it's any more contrived than the entirely background-dependent formulation of string theory remains to be seen.

The quantum theory of space-time in loop quantum gravity is really just a quantum theory of space. The spin network described by the theory cannot yet incorporate time. Some, such as Lee Smolin, believe that time will prove to be a necessary and fundamental component of the theory, while Carlo Rovelli believes that the theory will ultimately show that time doesn't really exist, but is just an emergent property without a real existence on its own. These and other disputes over the meaning of time are addressed in Chapter 16.

So Are These Two Theories the Same with Different Names?

One viewpoint is that both string theory and loop quantum gravity may actually represent the same theory approached from different directions. The parallels between the theories are numerous:

- ✔ String theory began as a theory of particle interactions, but was shown to contain gravity. Loop quantum gravity began as a theory of gravity, but was shown to contain particles.

- ✔ In string theory, space-time can be viewed as a mesh of interacting strings and branes, much like the threads of a fabric. In loop quantum gravity, threads of space are woven together, creating the apparently "smooth" fabric of space-time.

- ✔ Some string theorists believe the compactified dimensions represent a fundamental quantum unit of space, while LQG starts with units of space as an initial requirement.

- ✔ Both theories (provided certain assumptions are made) calculate the same entropy for black holes.

One way to view the differences is that string theory, which began by applying principles from particle physics, may point toward a universe in which space-time emerges from the behavior of these fundamental strings. LQG, on the other hand, began by applying general relativity principles and results in a world where space-time is fundamental, but matter and gravity may emerge from the behavior of these fundamental units.

At one time, Lee Smolin was one of the major supporters of the viewpoint that string theory, M-theory, and loop quantum gravity were different approximations of the same underlying fundamental theory. Over the last decade, he has become largely disillusioned with string theory (at least compared to his earlier conciliatory stance), becoming a prominent advocate of pursuing other avenues of inquiry.

Some string theorists believe that the methods used by LQG will eventually be carried over to string theory, allowing for a background-independent version of string theory. This is very probable, especially given that the string theory landscape seems capable of absorbing virtually any viable theory and incorporating it as a part of string theory.

Despite the possible harmony between the two fields, at the moment they are competitors for research funding and attention. String theorists have their conferences, and loop quantum gravity people have their conferences, and rarely shall the two conferences meet. (Except for Lee Smolin, who seems to have rather enjoyed flitting over to the string theory side of things over the years.) All too often, the groups seem unable to speak to each other in any meaningful way (see the nearby sidebar, "The 'Big Bang' breakup").

Part of the problem is one of sociology. Many string theorists, even in research papers, use phrases that make it clear they consider string theory to not only be their preferred theory, but to be the only (or, in cases where they're being more generous, the "most promising") theory of quantum gravity. By doing this, they often dismiss LQG as even being an option. Some string theorists have indicated in interviews that they are completely unaware of any viable alternatives to string theory! (This is because string theorists aren't yet convinced that the alternatives are actually viable.)

Hopefully, these physicists will find a way to work together and use their results and techniques in ways that provide real insights into the nature of our own universe. But so far, loop quantum gravity, like string theory, is still stuck on the drawing board.

The "Big Bang" breakup

The conflict between loop quantum gravity and string theory enthusiasts made it into popular culture in an episode of the CBS television sitcom *The Big Bang Theory,* which focuses on two physicist roommates, Leonard and Sheldon. In the second episode of the second season, Leonard has begun a relationship with a physicist colleague, Leslie Winkle, a rival of Sheldon (or "nemesis," as he thinks of her). Leslie Winkle, you see, is a researcher in loop quantum gravity, while Sheldon is a string theorist.

In the climactic scene of the episode, Sheldon and Leslie get into a "string war" of their own, slinging theoretical physics barbs at each other. Their conflict is over which theory — loop quantum gravity or string theory — has the greatest probability of successfully achieving a quantum theory of gravity. The argument ends in Leonard being placed in the middle, being forced to point out that they are two untested theories of quantum gravity, so he has no way to choose. Leslie is shocked and appalled by this response, immediately ending the relationship with Leonard.

Although this is obviously played up for comedy purposes, among the physics community the funniest thing about it was how much truth there actually was in the scenario. When physicists get into passionate debates about loop quantum gravity versus string theory, all too often the first casualty seems to be reasonable discourse.

Chapter 19

Considering Other Ways to Explain the Universe

In This Chapter

▶ Some physicists are working in areas other than string theory — honest!

▶ Working around the need for a theory of quantum gravity

▶ Seeking new mathematical approaches while solving string theory problems

*I*n the event that string theory proves false, or that there is no "theory of everything" at all, there are still some unexplained phenomena in the universe that require explanation. These issues mostly lie in the realm of cosmology, such as the flatness problem, dark matter, dark energy, and the details of the early universe.

Even though string theory is currently the dominant path being explored to answer most of these problems, some physicists have begun looking in other directions, beyond the loop quantum gravity described in Chapter 18. These rebels (and, at times, outcasts) have refused, in many cases, to stick with the mainstream theoretical community in adopting the principles of string theory and have proposed new directions of inquiry that are, at times, extremely radical — though possibly no more radical, in their own ways, than string theory was in the 1970s.

In this chapter, I explain some of the alternative approaches that physicists are looking into in an effort to explain the problems that physicists want to resolve. First, I explore some alternate quantum gravity theories, none of which are quite as fully developed as either string theory or loop quantum gravity. Next, I show you how physicists have suggested modifying the existing law of general relativity to take into account the facts that don't fit with Einstein's original model. It's possible that some of the ideas from this chapter will ultimately be incorporated into string theory, or perhaps take its place entirely.

Taking Other Roads to Quantum Gravity

Though string theorists like to point out that theirs is the most developed theory to unite general relativity and quantum physics (at times they even seem clueless that alternatives exist), sometimes it seems like nearly every physicist has come up with some plan to combine the two — they just don't have the support that string theorists have.

Most of these alternate theories start with the same idea as loop quantum gravity — that space is made up of small, discrete units that somehow work together to provide the space-time that we all know and love (relatively speaking, that is). Despite the fact that scientists don't know much about these units of space, some theorists can analyze how they might behave and use that information to generate useful models.

Here are some examples of these other quantum gravity approaches:

- ✔ **Causal dynamical triangulations (CDT):** CDT models space-time as being made up of tiny building blocks, called *4-simplices,* which are identical and can reconfigure themselves into different curvature configurations.

- ✔ **Quantum Einstein gravity (or "asymptotic safety"):** Quantum Einstein gravity assumes that there's a point where "zooming in" on space-time stops increasing the force of gravity.

- ✔ **Quantum graphity:** In the quantum graphity model, gravity didn't exist in the earliest moments of the universe because space itself doesn't exist on the small length and high energy scales involved in the early universe.

- ✔ **Internal relativity:** This model predicts that you can start with a random distribution of quantum spins and get the laws of general relativity to come out of it.

Of course, any of these approaches could advance either string theory or loop quantum gravity, instead of leading off in a new direction. Some of the principles may prove fruitful, but only when applied in the framework of one of the other theories. Only time will tell what insights, if any, come out of them and if they can be applied to give meaningful results.

Causal dynamical triangulations (CDT): If you've got the time, I've got the space

The causal dynamical triangulations approach consists of taking tiny building blocks of space, called *4-simplices* (sort of like multidimensional triangles), and using them to construct the space-time geometry. The result is a sequence of geometric patterns that are causally related in a sequence where one construction follows another (in other words, one pattern causes the next pattern). This system was developed by Renate Loll of Utrecht University in the Netherlands, and also by colleagues Jan Ambjørn and Jerzy Jurkiewicz.

One of the most important aspects of CDT is that time becomes an essential component of space-time, because Loll includes the causal link as a crucial part of the theory. Relativity tells us that time is distinctly different from space (as mentioned in Chapter 13, the time dimension has a negative in front of it in relativity), but Stephen Hawking and others have suggested that the difference between time and space could perhaps be ignored.

Loll then takes her causally linked configurations of 4-simplices and sums over all possible configurations of the shapes. (Feynman used a similar approach in quantum mechanics, summing over all possible paths to obtain quantum physics results.) The result is classical space-time geometry!

If true, CDT shows that it's impossible to ignore the difference between space and time. The causal link of changes in space-time geometry — in other words, the "time" part of space-time — is absolutely necessary to get classical space-time geometry that is governed by general relativity and matches what science knows of standard cosmological models.

At the tiniest scales, though, CDT shows that space-time is only 2-dimensional. The model turns into a fractal pattern, where the structures repeat themselves at smaller and smaller scales, and there's no proof that real space-time behaves that way.

CDT's biggest flaw in comparison to string theory is that it doesn't tell us anything about where matter comes from, whereas matter arises naturally in string theory from the interactions of fundamental strings.

Quantum Einstein gravity: Too small to tug

Quantum Einstein gravity, developed by Martin Reuter of the University of Mainz in Germany, tries to apply the quantum physics processes that worked on other forces to gravity. Reuter believes that at small scales, gravity may have a cutoff point where its strength stops increasing. (This notion was proposed by Steven Weinberg in the 1970s, under the more common name "asymptotic safety.")

One reason to think that gravity stops increasing at small scales is that this is what quantum field theory tells us the other forces do. At very small scales, even the strong nuclear force drops to zero. This is called *asymptotic freedom,* and its discovery earned David Gross, David Politzer, and Frank Wilczek the 2004 Nobel Prize. The force of gravity wouldn't go to zero but rather to some finite strength (stronger than we usually see), and this idea is known as asymptotic safety.

Weinberg and others weren't able to pursue the idea at the time because the mathematical tools to calculate the cutoff point for gravity in general relativity didn't exist until Reuter developed them in the 1990s. Though the method is approximate, Reuter has a great deal of confidence.

Quantum Einstein gravity, like CDT, comes up with a fractal pattern to small-scale space-time, and the number of dimensions drops to two. Reuter himself has noted that this could mean that his approach is fundamentally equivalent to CDT, because they both have these rather distinctive predictions at small scales.

The idea of asymptotic safety is really a very conservative solution to the problem of quantum gravity. Unlike the other approaches that introduce some radically new physics that would take over from general relativity at high energies (or equivalently at short distances), it proposes a well-defined strongly interacting theory of gravity at high energies in which the usual general relativity is simply augmented by some extra interactions for the graviton.

Quantum graphity: Disconnecting nodes

Quantum graphity has been developed by Fotini Markopoulou of the Perimeter Institute. In some ways, this is loop quantum gravity taken to its extreme — at extremely high energies all that exists is the network of nodes.

This model is based on a suggestion by John Archibald Wheeler about a *pre-geometric phase* to the universe, which Markopoulou takes literally. The nodes in the pre-geometric phase would all touch each other, but as the universe cooled, they would disconnect from each other and become separated, resulting in the space that we see today. (Physicists working on string theory have also found this sort of pre-geometric phase, so it's not unique to Markopoulou's approach.)

It's also possible that this could explain the horizon problem, the problem that distant parts of the universe seem to be the same temperature. In the quantum graphity model, all points used to be in direct contact, so inflation proves to be unnecessary. (See Chapter 9 for more about the horizon problem and how inflation solves it.) At present, inflation is a much more well-defined theory, but Markopoulou is working on developing quantum graphity to compete with it.

The Perimeter Institute

If you follow theoretical physics, it isn't long until you hear about the Perimeter Institute for Theoretical Physics, located in Waterloo, Ontario, Canada. The Perimeter Institute was founded in 1999 by Mike Lazaridis, who was founder and co-CEO of Research in Motion, the makers of the BlackBerry handheld device. Lazaridis decided to help foster research and innovation in Canada by starting the Perimeter Institute, which is devoted purely to theoretical physics research.

Many of the prominent critics of string theory who are working on other approaches — Lee Smolin, John Moffat, Fotini Markopoulou, and others — call it home, so it's easy to believe that the Perimeter Institute seeks out anti-string theorists. In fact, their current director is Neil

Turok, a cosmologist and co-creator of the ekpyrotic model, which is based on string theory principles. The Perimeter Institute achieved quite a coup by hiring Stephen Hawking as a Distinguished Research Chair, followed by a slew of other prominent physicists.

The Perimeter Institute's goal is to foster innovation, and the physicists work in a number of areas: cosmology, particle physics, quantum foundations, quantum gravity, quantum information theory, and superstring theory. It's one of the only places where string theorists and leaders in other quantum gravity approaches regularly work together in one institute. More information on the Perimeter Institute can be found at www.perimeter institute.ca.

Internal relativity: Spinning the universe into existence

The final quantum gravity model, internal relativity, may be the most ambitious, because Olaf Dreyer of MIT believes that a random distribution of quantum spins may end up resulting in our whole universe. For this to work, Dreyer considers the view of observers inside the system. The approach has shown that these observers would witness some aspects of special relativity, such as time dilation and length contraction, but Dreyer is still working on getting general relativity out of the equations. (Isn't everybody?)

The space-time and matter are a result of the excitations of the system, which is one reason Dreyer is hopeful. He believes that the reason quantum physics yields an incorrect prediction for the cosmological constant is because of a split between space-time and matter. Internal relativity links the two concepts, so the calculations have to be performed differently.

Dreyer has predicted that his model would show no gravity waves in the cosmic microwave background radiation (CMBR), while inflation theory would result in CMBR gravity waves. It is hoped that the Planck satellite will be able to detect any gravity waves in the CMBR — or not detect them, as Dreyer's theory predicts.

Newton and Einstein Don't Make All the Rules: Modifying the Law of Gravity

Instead of trying to develop theories of quantum gravity, some physicists are looking at the existing law of gravity and trying to find specific modifications that will make it work to explain the current mysteries of cosmology. These efforts are largely motivated by attempts to find alternatives to the cosmological theories of inflation, dark matter, or dark energy.

These approaches don't necessarily resolve the conflicts between quantum physics and general relativity, but in many cases they make the conflict less important. The approaches tend to result in singularities and infinities falling out of the theories, so there just isn't as much need for a theory of quantum gravity.

Doubly special relativity (DSR): Twice as many limits as ordinary relativity

One intriguing approach is *doubly special relativity* or *deformed special relativity* (abbreviated as DSR either way you slice it), originally developed by Giovanni Amelino-Camelia. In special relativity, the speed of light is constant for all observers. In DSR theories, all observers also agree on one other thing — the distance of the Planck length.

In Einstein's relativity, the constancy of the speed of light places an upper speed limit on everything in the universe. In DSR theories, the Planck length represents a lower limit on distance. Nothing can go faster than the speed of light, and nothing can be smaller than a Planck length. The principles of DSR may be applicable to various quantum gravity models, such as loop quantum gravity, though so far there's no proof for it.

Modified Newtonian dynamics (MOND): Disregarding dark matter

Some physicists aren't comfortable with the idea of dark matter and have proposed alternative explanations to resolve the problems that make physicists believe dark matter exists. One of these explanations, which involves looking at gravity in a new way on large scales, is called *modified Newtonian dynamics (MOND)*.

The basic premise of MOND is that at low values, the force of gravity doesn't follow the rules laid out by Newton more than 300 years ago. The relationship between force and acceleration in these cases may turn out not to be exactly linear, and MOND predicts a relationship that will yield the results observed based on only the visible mass for galaxies.

In Newtonian mechanics (or, for that matter, in general relativity, which reduces to Newtonian mechanics at this scale), the gravitational relationships between objects are precisely defined based on their masses and the distance between them. When the amount of visible matter for galaxies is put into these equations, physicists get answers that show that the visible matter just doesn't produce enough gravity to hold the galaxies together. In fact, according to Newtonian mechanics, the outer edges of the galaxies should be rotating much faster, causing the stars farther out to fly away from the galaxy.

Because scientists know the distances involved, the assumption is that somehow the amount of matter has been underestimated. A natural response to this (and the one that most physicists have adopted) is that there must be some other sort of matter that isn't visible to us: dark matter.

There is one other alternative — the distances and matter are correct, but the relationship between them is incorrect. MOND was proposed by Israeli physicist Mordehai Milgrom in 1981 as a means of explaining the galactic behavior without resorting to dark matter.

Most physicists have ruled MOND out, because the dark matter theories seem to fit the facts more closely. Milgrom, however, has not given up, and in 2009 he made predictions about slight variations in the path of planets based on his MOND calculations. It remains to be seen if these variations will be observed.

Variable speed of light (VSL): Light used to travel even faster

In two separate efforts, physicists have developed a system where the speed of light actually would not be constant, as a means of explaining the horizon problem without the need of inflation. The earliest system of the *variable speed of light (VSL)* was proposed by John Moffat (who later incorporated the idea into his modified gravity theory), and a later system was developed by João Magueijo and Andreas Albrecht.

Proving dark matter wrong?

In August 2008, a group of astrophysicists published a paper called "A Direct Empirical Proof of the Existence of Dark Matter." The "proof" they speak of came from an impact between two galaxy clusters. Using NASA's Chandra X-Ray Observatory, they were able to see *gravitational lensing* (the gravity of the collision caused light to bend, kind of how light bends when it passes through a lens), which let them determine the center of the collision. The center of the collision did *not* match the center of the visible matter. In other words, the center of gravity and the center of visible matter didn't match. That's pretty conclusive evidence for there being nonvisible matter, right?

In the world of theoretical physics, nothing is quite that easy these days. By September, physicist John Moffat and others were beginning to cast doubt on whether dark matter was the only explanation. Using his own *modified gravity (MOG)* theory, Moffat performed a calculation on a simplified 1-dimensional version of the collision.

Most physicists accept the NASA findings, including more recent findings from WMAP and other observations, as conclusive evidence that dark matter exists. But there remain those who are unconvinced and search for other explanations.

The horizon problem is based on the idea that distant regions of the universe couldn't communicate their temperatures because they are so far apart light hasn't had time to get from one to the other. The solution proposed by inflation theory is that the regions were once much closer together, so they could communicate (see Chapter 9 for more on this).

In VSL theories, another alternative is proposed: The two regions could communicate because light traveled faster in the past than it does now.

Moffat proposed his VSL model in 1992, allowing for the speed of light in the early universe to be very large — about 100,000 trillion trillion times the current values. This would allow for all regions of the observable universe to easily communicate with each other.

To get this to work out, Moffat had to make a conjecture that the *Lorentz invariance* — the basic symmetry of special relativity — was somehow spontaneously broken in the early universe. Moffat's prediction results in a period of rapid heat transfer throughout the universe that results in the same effects as an inflationary model.

In 1998, physicist João Magueijo came up with a similar theory, in collaboration with Aldreas Albrecht. Their approach, developed without any knowledge of Moffat's work, was very similar — which they acknowledged upon learning of it. This work was published a bit more prominently than Moffat's (largely because they were more stubborn about pursuing publication in the prestigious *Physical Review D,* which had rejected Moffat's earlier paper). This later work has inspired others, such as Cambridge physicist John Barrow, to begin investigating this idea.

One piece of support for VSL approaches is that recent research by John Webb and others has indicated that the fine-structure constant may not have always been constant. The *fine-structure constant* is a ratio made up from Planck's constant, the charge on the electron, and the speed of light. It's a value that shows up in some physical equations. If the fine-structure constant has changed over time, then at least one of these values (and possibly more than one) has also been changing.

The spectral lines emitted by atoms are defined by Planck's constant. Scientists know from observations that these spectral lines haven't changed, so it's unlikely that Planck's constant has changed. (Thanks to John Moffat for clearing that up.) Still, any change in the fine-structure constant could be explained by varying either the speed of light or the electron charge (or both).

Physicists Elias Kiritsis and Stephon Alexander independently developed VSL models that could be incorporated into string theory, and Alexander later worked with Magueijo on refining these concepts (even though Magueijo is critical of string theory's lack of contact with experiment).

These proposals are intriguing, but the physics community in general remains committed to the inflation model. Both VSL and inflation require some strange behavior in the early moments of the universe, but it's unclear that inflation is inherently more realistic than VSL. It's possible that further evidence of varying constants will ultimately lead to support of VSL over inflation, but that day seems a long way off, if it ever happens.

Modified gravity (MOG): The bigger the distance, the greater the gravity

John Moffat's work in alternative gravity has resulted in his *modified gravity (MOG) theories,* in which the force of gravity increases over distance, and also the introduction of a new repulsive force at even larger distances. Moffat's MOG actually consists of three different theories that he has developed over the span of three decades, trying to make them simpler and more elegant and more accessible for other physicists to work on.

This work began in 1979, when Moffat developed *nonsymmetric gravitational theory (NGT),* which extended work that Einstein tried to apply to create a unified field theory in the context of a non-Riemannian geometry. The work had failed to unify gravity and electromagnetics, like Einstein wanted, but Moffat believed that it could be used to generalize relativity itself.

Over the years, NGT ultimately proved inconclusive. It was possible that its predictions (such as the idea that the sun deviated from a perfectly spherical shape) was incorrect or that the deviation was too small to be observed.

In 2003, Moffat developed an alternative with the unwieldy name *Metric-Skew-Tensor Gravity (MSTG)*. This was a symmetric theory (easier to deal with), which included a "skew" field for the nonsymmetric part. This new field was, in fact, a fundamentally new force — a fifth fundamental force in the universe.

Unfortunately, this theory remained too mathematically complicated in the eyes of many, so in 2004 Moffat developed *Scalar-Tensor-Vector Gravity (STVG)*. In STVG, Moffat again had a fifth force resulting from a vector field called a *phion field*. The phion particle was the gauge boson that carried the fifth force in the theory.

According to Moffat, all three theories give essentially the same results for weak gravity fields, like those we normally observe. The strong gravitational fields needed to distinguish the theories are the ones that always give scientists problems and have motivated the search for quantum gravity theories in the first place. They can be found at the moment of the big bang or during the stellar collapses that may cause black holes.

There are indications that STVG yields results very similar to Milgrom's MOND theory (refer to the earlier section "Modified Newtonian dynamics (MOND): Disregarding dark matter" for a fuller explanation of MOND). Moffat has proposed that MOG may actually explain dark matter and dark energy, and that black holes may not actually exist in nature.

While these implications are amazing, the work is still in the very preliminary stages, and it will likely be years before it (or any of the other theories) is developed enough to have any hope of seriously competing with the entrenched viewpoints.

Rewriting the Math Books and Physics Books at the Same Time

Revolutions in physics have frequently had an assist from revolutions in mathematics years before. One of the problems with string theory is that it has advanced so quickly that the mathematical tools didn't actually exist. Physicists have been forced (with the aid of some brilliant mathematicians) to develop the tools as they go.

Einstein got help in developing general relativity from Riemannian geometry, developed years earlier. Quantum physics was built on a framework of new mathematical representations of physical symmetries, group representation theory, as developed by the mathematician Hermann Weyl.

In addition to developing the physics needed to address problems of quantum gravity, some physicists and mathematicians have tried to focus on developing whole new mathematical techniques. The question remains, though, how (and if) these techniques could be applied to the theoretical frameworks to get meaningful results.

Compute this: Quantum information theory

One technique that is growing in popularity as a means of looking at the universe is *quantum information theory,* which deals with all elements in the universe as pieces of information. This approach was originally proposed by John Archibald Wheeler with the phrase "It from bit," indicating that all matter in the universe can be viewed as essentially pieces of information. (A *bit* is a unit of information stored in a computer.)

Some of the leaders in this approach are Fotini Markopoulou of the Perimeter Institute and Seth Lloyd of MIT, who approach the problem from rather different directions. Markopoulou studies quantum gravity theories, while Lloyd is best known for having figured out how to build a *quantum computer.* (Quantum computers are like ordinary computers, but instead of using just two bits for information storage, they use quantum physics to have a whole host of in-between information. A quantum bit of information is called a *qubit.*)

Overall, this approach basically treats the universe as a giant computer — in fact, a universe-sized quantum computer. The major benefit of this system is that, for a computer scientist, it's easy to see how random information sent through a series of computations results in complexity growing over time. The complexity within our universe could thus arise from the universe performing logical operations — calculations, if you will — upon the pieces of information (be they loops of space-time or strings) within the universe.

If you want to know more about quantum information theory, or quantum computers for that matter, you can read about it in Seth Lloyd's 2006 physics book, *Programming the Universe: A Quantum Computer Scientist Takes on the Cosmos,* which should be accessible if you've followed the science in this book.

Looking at relationships: Twistor theory

For nearly four decades, the brilliant physicist Sir Roger Penrose has been exploring his own mathematical approach — *twistor theory*. Penrose developed the theory out of a strong general relativity approach (the theory requires only four dimensions). Penrose maintains a belief that any theory of quantum gravity will need to include fundamental revisions to the way physicists think about quantum mechanics, something with which most particle physicists and string theorists disagree.

One of the key aspects of twistor theory is that the relation between events in space-time is crucial. Instead of focusing on the events and their resulting relationships, twistor theory focuses on the causal relationships, and the events become byproducts of those relationships.

If you take all of the light rays in space-time, it creates a *twistor space,* which is the mathematical universe in which twistor theory resides. In fact, there are some indications that objects in twistor space may result in objects and events in our universe.

The major flaw of twistor theory is that even after all of these years (it was originally developed in the 1960s), it still only exists in a world absent of quantum physics. The space-time of twistor theory is perfectly smooth, so it allows no discrete structure of space-time. It's a sort of anti-quantum gravity, which means it doesn't provide much more help than general relativity in resolving the issues that string theorists (or other quantum gravity researchers) are trying to solve.

As with string theory, Penrose's twistor theory has provided some mathematical insights into the existing theories of physics, including some that lie at the heart of the Standard Model of particle physics.

Edward Witten and other string theorists have begun to investigate ways that twistor theory may relate to string theory. One approach has been to have the strings exist not in physical space, but in twistor space. So far, it hasn't yielded the relationships that would provide fundamental breakthroughs in either string theory or twistor theory, but it has resulted in great improvements of calculational techniques in quantum chromodynamics.

Uniting mathematical systems: Noncommutative geometry

Another mathematical tool being developed is the *noncommutative geometry* of French mathematician Alain Connes, a winner of the prestigious Fields Medal. This system involves treating the geometry in a fundamentally new way, using mathematical systems where the commutative principle doesn't hold.

In mathematics, two quantities *commute* if operations on those quantities work the same way no matter what order you treat them. Addition and multiplication are both commutative because you get the same answer no matter what order you add two numbers or multiply them.

However, mathematicians are a diverse bunch, and some mathematical systems exist where addition and multiplication are defined differently, so the order does matter. As weird as it sounds, in these systems multiplying 5 by 3 could give a different result than multiplying 3 by 5. (I don't recommend using this excuse to argue with a teacher over the scores on a math test.) It's probably not surprising to discover that these noncommutative mathematical systems come up frequently in the bizarre world of quantum mechanics — in fact, this feature is the mathematical cause of the uncertainty principle described in Chapter 7.

The tools of noncommutative geometry have been used in many approaches, but Connes seeks a more fundamental unification of algebra and geometry that could be used to build a physical model where the conflicts are resolved by features inherent in the mathematical system.

Noncommutative geometry has had some success, because the Standard Model of particle physics seems to pop out of it in the simplest versions. The goal of the committed mathematicians working with Connes is that they will eventually be able to replicate all of physics (including possibly string theory), though that is likely still a long way off. (Are you beginning to see a pattern here?)

Part VI
The Part of Tens

The 5th Wave   By Rich Tennant

In this part . . .

In these classic *For Dummies* Part of Tens chapters, I offer some greater insights into what string theory might accomplish and the people closest to it.

I explore ten concepts that physicists hope a "theory of everything" will explain, whether or not string theory turns out to be that theory.

I also give you some background on ten of the most prominent string theorists working to show that string theory is the way to unite quantum theory and general relativity.

Chapter 20

Ten Questions a Theory of Everything Should (Ideally) Answer

*A*ny "theory of everything" — whether it be string theory or something else — would need to answer some of the most difficult questions that physics has ever asked. These questions are so difficult that the combined efforts of the entire physics community have so far been unable to answer them. Most physicists have, historically, believed that a theory of everything would provide a unique reason why the universe is the way it is — as opposed to the anthropic principle, which is based on our universe *not* being unique. Many physicists today question whether there can ever be a single theory that answers all of these questions.

In this chapter, I consider the questions of what started the universe, including why the early universe had exactly the properties it had. This includes the solutions to other questions of cosmology, such as the nature of black holes, dark matter, and dark energy. I also explore the problem of understanding what really happens in the strange realm of quantum physics. Finally, I discuss the need for a fundamental explanation of time and a reasonable look ahead at the end of the universe.

The Big Bang: What Banged (and Inflated)?

Currently, physics and cosmology tell us that the universe as we know it started about 14 billion or so years ago, in a singularity at which the laws of physics break down. Most scientists believe in a rapid inflation that occurred moments afterward, expanding space rapidly. When the inflation period slowed down, we entered into a period where space continued to expand at the rate we see today (or a bit less, given dark energy's influences).

This breaks the question of the universe's origin into two parts:

✔ What were the initial conditions that triggered the big bang?

✔ What caused the repulsive gravity of the inflation era to end?

In Chapter 14, I offer some explanations for how string theory can solve these questions. Even if string theory fails, any theory that attempts to expand beyond the Standard Model of particle physics will need to tackle these questions regarding the early moments of the universe.

Baryon Asymmetry: Why Does Matter Exist?

After the big bang, raw energy was transformed into matter. If the energy of the early universe had cooled into equal amounts of matter and antimatter, these different forms of matter would have annihilated each other, leaving no matter in the universe. Instead, there was substantially more matter than antimatter, enough so that when all the antimatter had been annihilated by matter, enough matter was left to make up the visible universe. This early difference between matter and antimatter is called *baryon asymmetry* (because regular matter, made up of baryons, is called *baryonic matter*).

The laws of physics provide no clear reason why the amounts of matter and antimatter wouldn't have been equal, so presumably a theory of everything would explain why the dense energy of the early universe tended to favor — even if only by a little bit — matter over antimatter.

Hierarchy Issues: Why Are There Gaps in Forces, Particles, and Energy Levels?

Most physicists, if they were to set out to create a universe, would have been a bit more conservative with their resources than the forces at work in our universe seem to have been. There are a wide range of force intensities, ranging from the incredibly weak gravitational force to the strong nuclear force that binds quarks together into protons and neutrons. The particles themselves come in multiple varieties — far more varieties than we seem to need — and each variety jumps by large multiples in size. Instead of a smooth continuum of forces, particles, and energy, there are huge gaps.

A theory of everything should explain why these gaps exist and why they exist where they do.

Fine-Tuning: Why Do Fundamental Constants Have the Values They Do?

Many of the fundamental constants in our universe seem precisely set in the range that allows life to form. This is one reason why some physicists have been turning toward the anthropic principle, because it so readily explains this fact.

Physicists hope, however, that a theory of everything would explain the precision of these values — in essence, explain the reason why life itself is allowed to exist in our universe — from fundamental principles of physics.

Black Hole Information Paradox: What Happens to Missing Black Hole Matter?

The current thinking on the black hole information paradox is that there is a quantum system underlying the black hole, and that this quantum system never loses information, though the system can mix up the finer points in a complicated way. To reconcile this picture with Hawking's calculations (see Chapter 14), the concept of complementarity is sometimes invoked. This idea, proposed by Leonard Susskind, says that someone outside the black hole may observe different results than someone falling into the black hole, but that no contradictions will arise.

This approach hasn't settled the problem for everyone, including physicists who believe that relativity should hold more sway than quantum mechanics. Whatever the solution, a theory of everything would have to present a definitive set of rules that could be applied to figure out what's happening to matter (and information) that falls into a black hole.

Quantum Interpretation: What Does Quantum Mechanics Mean?

Though quantum mechanics works to explain the results seen in laboratory experiments, there's still not a single clear description of the physical principle that causes it to work the way it does. Though this is tied to the "collapse of the quantum wavefunction," the exact physical meaning of the wavefunction, or of its collapse, remains a bit of a mystery. (So if you don't understand quantum physics, don't worry . . . physicists are still debating it, even after all these years.)

In Chapter 7, I explain some of the interpretations of what this may mean — the Copenhagen interpretation, the many worlds interpretation (MWI), consistent histories, and so on — but the fact is that these are just guesses, and physicists really don't know for sure what's going on with this strange quantum behavior. Lee Smolin listed this as his second "great problem in theoretical physics." Though today this is by far the minority opinion among physicists, the great physicists of the quantum revolution — Bohr, Einstein, Heisenberg, Schroedinger, and the rest — also saw it as a key question to resolve.

Today, most physicists tend to just trust in the math and don't worry about strange things happening behind the scenes. They are perfectly comfortable with quantum mechanics, seeing nothing mysterious in the behavior. (After all, they have equations that describe it!)

In fact, the majority of theoretical physicists don't seem to believe that it's possible to determine one interpretation as correct, and don't even consider it as a question that needs to be answered, even by a theory of everything. Some of those who do want a clear interpretation hope that a theory of everything will provide insights into the physical mechanism explaining quantum phenomena.

Dark Mystery No. 1: What Is Dark Matter (and Why Is There So Much)?

There appear to be two forms of matter in the universe: visible matter and dark matter. Scientists know dark matter exists because they can detect its gravitational effects, but they can't currently observe dark matter directly. If extra matter weren't there to hold galaxies together, the equations of general relativity show that they would fly apart.

Still, no one knows what the dark matter is made of. Some theorize that the dark matter may be stable superpartners of our known particles — perhaps photinos, the superpartner of the photon. String theory contains other ideas, covered in Chapter 14, that could explain the nature of dark matter.

But the fact is that no one knows for sure, which is disturbing because there is about five times as much dark matter as there is visible matter in the universe. So there should be a lot of it around to study — if only physicists and their scientific theories could see it for what it really is.

Dark Mystery No. 2: What Is Dark Energy (and Why Is It So Weak)?

There's a lot of dark energy in the universe — about three times as much as visible matter and dark matter put together! This energy represents a repulsive force of gravity on large scales, pushing the edges of the universe apart.

The abundance of dark energy by itself isn't so much a problem; the real problem is that the dark energy is a lot weaker than physicists would expect from purely theoretical calculations based on quantum field theory. According to those calculations, the random energy of empty space (the "vacuum energy") should explode up to huge quantities, but instead it maintains an incredibly small value.

A theory of everything would hopefully explain why the vacuum energy contains the value it does.

Time Symmetry: Why Does Time Seem to Move Forward?

The space dimensions are interchangeable, but time is distinct because it seems to move in only one direction. This doesn't have to be the case. In fact, the mathematical laws of physics work either way, even in a universe where time could run backwards.

But time doesn't run backwards at all, and a theory of everything would need to explain this discrepancy between the mathematical symmetry of time and the physical asymmetry of time that you observe every time you're running late for a meeting.

The End of the Universe: What Comes Next?

And, of course, the eternal question of the fate of the universe is another question that a theory of everything would need to answer. (Cue up the song "It's the End of the World as We Know It" by R.E.M.) Will our universe (and all the others) end in ice, expanding until heat dissipates out across the vastness of space? Will galaxies huddle together in dense clusters, like winter campers around a campfire? Will the universe contract together and perhaps eventually start the cycle of universal creation all over?

Chances are that these questions will be answered long after we're gone, but there is hope that the beginnings of those answers may come within the next few years, as some aspects of string theory begin to enter the realm of experimental verification.

Chapter 21

Ten Notable String Theorists

*N*o new theory can develop without dedicated scientists working hard to refine and interpret it. Throughout this book, you read about some of the pioneering work in string theory. Now it's time to find out more about some of the scientists themselves, the people who make string theory tick as they research the mysteries of the universe within the context of this budding science. As string theory unfolds, some of these individuals may become legends on the order of Einstein and Newton, or they may end up finding useful ways of presenting this complex theory in ways that the general public can understand.

In this chapter, I introduce ten physicists who are responsible for the rise of string theory. I give brief biographies of not only the founders of string theory, but also some of the visionaries who have refined the theory over the years. Some of these personalities are also physicists who have written popular books or been involved with educational programs on the topic, helping to broaden the general public's understanding of string theory. However, this chapter isn't a "top ten" list, and just because a name hasn't been included should not be taken to mean that the person's work and contributions are any less significant than the names listed.

Edward Witten

Seen by many as the leading thinker of string theory, Witten introduced the concept of M-theory in 1995 as a way to consolidate the existing string theories into one comprehensive theory. Witten's work in string theory also included the 1984 application of Calabi-Yau manifolds to explain the compactification of the extra dimensions.

In 1951, Witten was born into physics, in a sense; his father, Louis Witten, was a theoretical physicist specializing in general relativity. Growing up, Witten displayed a natural aptitude for mathematics. Despite this, he focused his early years on studying history and being politically active, helping with George McGovern's 1972 presidential campaign. His undergraduate degree from Brandeis University was in history with a minor in linguistics.

In the fall of 1973, Witten went to graduate school in applied mathematics at Princeton University. Despite lacking a physics undergraduate degree, he quickly showed himself to be proficient at the complex mathematics involved in theoretical physics. He switched to the physics department and received his PhD from Princeton in 1976.

Witten has since published more than 300 research papers. According to some sources, he has the largest *h-index* (most often cited papers) of any living physicist. He received the MacArthur Foundation "genius grant" fellowship in 1982. In 1990, he was the first (and so far only) physicist to receive the Fields Medal, sometimes informally called the "Nobel Prize of mathematics" (the Nobel Committee awards no mathematics prize). He was one of *Time* magazine's 100 most influential people in 2004.

Among string theorists, Edward Witten is seen as a guiding light because of his ability to grasp the implications of the complex mathematics of the theory on a level that few others have been able to match. Even the strongest string theory critics speak of his intellect and mathematical prowess in awe, making clear that he is an unparalleled mind of his generation.

John Henry Schwarz

If string theory were a religion, then John Henry Schwarz would be the equivalent of St. Paul. At a time when virtually every other physicist abandoned string theory, Schwarz persevered for almost a decade as one of the few who tried to work out the theory's mathematical details, even though it hurt his career. Eventually, his work led to the first superstring revolution.

Schwarz was one of the physicists who discovered that supersymmetry resolved several of the problems with string theory. Later, Schwarz proposed the idea that the spin-2 particle described by string theory may be the graviton, meaning that string theory could be the long-sought theory to unify quantum physics and general relativity. (See Chapter 10 for more on these concepts.)

Schwarz worked at Caltech for 12 years — from 1972 to 1984 — as a temporary researcher instead of a full professor. His career prospects were hindered in large part because of his perceived obsession with string theory.

In 1984, Schwarz performed (along with Michael Green) the work showing that string theory was consistent, triggering the first superstring revolution. Without Schwarz's decade of dedicated work (or obsession), there would have been no foundation in place for superstring theory to build upon throughout the 1980s, when it rose to prominence among particle physicists.

Yoichiro Nambu

Yoichiro Nambu is one of the founders of string theory who independently discovered the physical description of the Veneziano model as vibrating strings. Nambu was already a respected particle physicist for his earlier work in describing the mechanism of spontaneous symmetry breaking in particle physics. Dr. Nambu received the 2008 Nobel Prize in Physics for this work.

Though this makes him the only founder of string theory to have received a Nobel Prize, it's important to note that the Nobel award makes no mention of string theory. In fact, the Nobel can't be awarded for theoretical work that hasn't been confirmed or proved useful experimentally.

Leonard Susskind

Leonard Susskind is another founder of string theory. As he recounts in his book *The Cosmic Landscape: String Theory and the Illusion of Intelligent Design,* he saw the original dual resonance model equations and thought they looked similar to equations for oscillators, which led him to create the string description — concurrently with Yoichiro Nambu and Holger Nielson. In addition, he has proposed several concepts discussed throughout this book: string theory of black hole entropy (Chapter 14), the holographic principle (Chapter 11), matrix theory (Chapter 11), and the application of the anthropic principle to the string theory landscape (which is the subject of *The Cosmic Landscape;* I cover this principle in Chapter 11).

In addition to his extensive work in string theory, Susskind is well-known for his disagreements with Stephen Hawking over the final fate of information that falls into a black hole, as outlined in his 2008 book *The Black Hole War: My Battle with Stephen Hawking to Make the World Safe for Quantum Mechanics.*

David Gross

David Gross was one of the physicists who developed the heterotic string theory, one of the major findings of the first superstring revolution.

In 2004, Gross earned (along with colleagues Frank Wilczek and David Politzer) the Nobel Prize in Physics for their 1973 discovery of asymptotic freedom in the strong nuclear interaction of quarks. (This means that the strong interaction between quarks gets weaker at extremely short distances.)

Since 1997, Dr. Gross has been the director of the Kavli Institute for Theoretical Physics at the University of California, Santa Barbara. In this capacity, Gross is known not only as a strong advocate for string theory but also as a strong opponent of the anthropic principle as applied to the string theory landscape.

Joe Polchinski

Joe Polchinski proved that string theory required objects of more than one dimension, called branes. Although the concept of branes had previously been introduced, Polchinski explored the nature of D-branes. This work was crucial to the second superstring revolution of 1995. Polchinski's work is seen as fundamental to the development of M-theory, brane world scenarios, and the holographic principle (all covered in Chapter 11).

Lately, Polchinski has become a convert to the anthropic principle's usefulness in string theory, though stories abound of how he once loathed the principle, considering it unscientific and threatening to quit his position if he were forced to adopt it.

Juan Maldacena

Juan Maldacena is an Argentine physicist who developed the idea that a duality exists between string theory and a quantum field theory — called the Malcadena duality (or the AdS/CFT correspondence; see Chapter 11).

The Maldacena duality, proposed in 1997, has been applied only in certain cases, but if it can be extended to all of string theory, it would allow a means to give a precise quantum string theory. In other words, string theorists should be able to translate known principles of gauge field theory into string theory equations — an excellent starting point for a complete quantum theory of gravity. Also, applying the duality in the other direction, starting with string theory and creating predictions about how gauge theory should behave could yield predictions that are testable at the Relativistic Heavy Ion Collider or Large Hadron Collider in years to come.

Lisa Randall

Theoretical physics is a realm stereotypically dominated by men and, even among the rare women who choose it, Lisa Randall doesn't fit the mold. She spends her free time on intense rock climbing expeditions but spends her professional days exploring the implications of multidimensional brane worlds as a phenomenologist.

Dr. Randall was the first tenured woman in the physics department at Princeton University. She was also the first tenured female theoretical physicist at MIT and later at Harvard, where she has been since 2001.

Randall rose to prominence among nonphysicists with her 2005 book *Warped Passages: Unraveling the Mysteries of the Universe's Hidden Dimensions*. Among other things, this resulted in her appearance on Comedy Central's wildly popular mock political pundit show, *The Colbert Report*.

Given her success as a woman in a male-dominated field, it's not surprising that she has impressive credentials. One of the most intriguing models to come out of her analysis of brane world scenarios are the Randall-Sundrum models, which explore the possibility of gravity behaving differently off of our own 3-brane.

Michio Kaku

Physicist Michio Kaku has been one of the most vocal supporters of string theory. He worked on the theory early in the 1970s, actually co-founding "string field theory" by writing string theory in a field form. By his own account, he then abandoned work on string theory because he didn't believe in the additional dimensions the theory demanded. He returned to string theory during the first superstring revolution and has proven an entertaining and lucid spokesman ever since.

Dr. Kaku wrote one of the first popular books on the topic, *Hyperspace: A Scientific Odyssey Through Parallel Universes, Time Warps, and the 10th Dimension,* in 1994. (This was my first introduction to string theory, when I read the book as a high school senior.) He has since written other books on futurism and advanced scientific and technology principles. His 2005 book, *Parallel Worlds,* focuses on many topics related to string theory.

For more than 25 years, Kaku has been a professor of theoretical physics at the City College of New York. The close proximity to major television networks may explain why he regularly appears on so many television programs. With a distinctive mane of white hair, Dr. Kaku is easily recognizable when he makes appearances on CNN, Discovery, the Science Channel, or ABC's *Good Morning, America.* (When *GMA* needed someone to explain how Mentos cause soda bottles to erupt into fountains of fizz, they called in Dr. Kaku.)

Kaku has also hosted a number of programs, including two of his own radio shows. He is currently seen hosting the SciQ Sunday specials on the Science Channel. His research work on the subject of string theory isn't as impressive as the others on this list, but he has done an incredible amount to popularize the ideas of string theory. Many recognize him as one of the theory's most vocal proponents to layman audiences.

Brian Greene

Last but certainly not least is probably one of the best-known string theorists, especially among nonphysicists. Brian Greene's popularity as a writer and spokesman for the field dates back to his 1999 book *The Elegant Universe: Superstrings, Hidden Dimensions, and the Quest for the Ultimate Theory,* which was used in 2003 as the basis for a three-part PBS *Nova* special. In 2004, Greene followed up with the book *The Fabric of the Cosmos: Space, Time, and the Fabric of Reality.* (He has appeared on Comedy Central's *The Colbert Report* at least twice, outdoing Dr. Randall's one appearance.)

Dr. Greene earned his undergraduate degree from Harvard. As a Rhodes Scholar, he received a 1986 doctorate from Oxford University. He was a professor at Cornell University for several years, but has been a full professor at Columbia University since 1996. Throughout his career, his research has focused on quantum geometry and attempting to understand the physical meaning of the extra dimensions implied by string theory.

In addition to trying to explain string theory to the masses, Greene has been co-director of Columbia University's Institute for Strings, Cosmology and Astroparticle Physics (ISCAP) since its founding in 2000.

In 2008, Greene was a founder of the World Science Festival in New York City, where a dance troupe performed an interpretative dance version of his book *The Elegant Universe.*

Index

Business/Accounting & Bookkeeping

Bookkeeping For Dummies
978-0-7645-9848-7

eBay Business
All-in-One For Dummies,
2nd Edition
978-0-470-38536-4

Job Interviews
For Dummies,
3rd Edition
978-0-470-17748-8

Resumes For Dummies,
5th Edition
978-0-470-08037-5

Stock Investing
For Dummies,
3rd Edition
978-0-470-40114-9

Successful Time
Management
For Dummies
978-0-470-29034-7

Computer Hardware

BlackBerry For Dummies,
3rd Edition
978-0-470-45762-7

Computers For Seniors
For Dummies
978-0-470-24055-7

iPhone For Dummies,
2nd Edition
978-0-470-42342-4

Laptops For Dummies,
3rd Edition
978-0-470-27759-1

Macs For Dummies,
10th Edition
978-0-470-27817-8

Cooking & Entertaining

Cooking Basics
For Dummies,
3rd Edition
978-0-7645-7206-7

Wine For Dummies,
4th Edition
978-0-470-04579-4

Diet & Nutrition

Dieting For Dummies,
2nd Edition
978-0-7645-4149-0

Nutrition For Dummies,
4th Edition
978-0-471-79868-2

Weight Training
For Dummies,
3rd Edition
978-0-471-76845-6

Digital Photography

Digital Photography
For Dummies,
6th Edition
978-0-470-25074-7

Photoshop Elements 7
For Dummies
978-0-470-39700-8

Gardening

Gardening Basics
For Dummies
978-0-470-03749-2

Organic Gardening
For Dummies,
2nd Edition
978-0-470-43067-5

Green/Sustainable

Green Building
& Remodeling
For Dummies
978-0-470-17559-0

Green Cleaning
For Dummies
978-0-470-39106-8

Green IT For Dummies
978-0-470-38688-0

Health

Diabetes For Dummies,
3rd Edition
978-0-470-27086-8

Food Allergies
For Dummies
978-0-470-09584-3

Living Gluten-Free
For Dummies
978-0-471-77383-2

Hobbies/General

Chess For Dummies,
2nd Edition
978-0-7645-8404-6

Drawing For Dummies
978-0-7645-5476-6

Knitting For Dummies,
2nd Edition
978-0-470-28747-7

Organizing For Dummies
978-0-7645-5300-4

SuDoku For Dummies
978-0-470-01892-7

Home Improvement

Energy Efficient Homes
For Dummies
978-0-470-37602-7

Home Theater
For Dummies,
3rd Edition
978-0-470-41189-6

Living the Country Lifestyle
All-in-One For Dummies
978-0-470-43061-3

Solar Power Your Home
For Dummies
978-0-470-17569-9

Internet
Blogging For Dummies,
2nd Edition
978-0-470-23017-6

eBay For Dummies,
6th Edition
978-0-470-49741-8

Facebook For Dummies
978-0-470-26273-3

Google Blogger
For Dummies
978-0-470-40742-4

Web Marketing
For Dummies,
2nd Edition
978-0-470-37181-7

WordPress For Dummies,
2nd Edition
978-0-470-40296-2

Language & Foreign Language
French For Dummies
978-0-7645-5193-2

Italian Phrases
For Dummies
978-0-7645-7203-6

Spanish For Dummies
978-0-7645-5194-9

Spanish For Dummies,
Audio Set
978-0-470-09585-0

Macintosh
Mac OS X Snow Leopard
For Dummies
978-0-470-43543-4

Math & Science
Algebra I For Dummies
978-0-7645-5325-7

Biology For Dummies
978-0-7645-5326-4

Calculus For Dummies
978-0-7645-2498-1

Chemistry For Dummies
978-0-7645-5430-8

Microsoft Office
Excel 2007 For Dummies
978-0-470-03737-9

Office 2007 All-in-One
Desk Reference
For Dummies
978-0-471-78279-7

Music
Guitar For Dummies,
2nd Edition
978-0-7645-9904-0

iPod & iTunes
For Dummies,
6th Edition
978-0-470-39062-7

Piano Exercises
For Dummies
978-0-470-38765-8

Parenting & Education
Parenting For Dummies,
2nd Edition
978-0-7645-5418-6

Type 1 Diabetes
For Dummies
978-0-470-17811-9

Pets
Cats For Dummies,
2nd Edition
978-0-7645-5275-5

Dog Training For Dummies,
2nd Edition
978-0-7645-8418-3

Puppies For Dummies,
2nd Edition
978-0-470-03717-1

Religion & Inspiration
The Bible For Dummies
978-0-7645-5296-0

Catholicism For Dummies
978-0-7645-5391-2

Women in the Bible
For Dummies
978-0-7645-8475-6

Self-Help & Relationship
Anger Management
For Dummies
978-0-470-03715-7

Overcoming Anxiety
For Dummies
978-0-7645-5447-6

Sports
Baseball For Dummies,
3rd Edition
978-0-7645-7537-2

Basketball For Dummies,
2nd Edition
978-0-7645-5248-9

Golf For Dummies,
3rd Edition
978-0-471-76871-5

Web Development
Web Design All-in-One
For Dummies
978-0-470-41796-6

Windows Vista
Windows Vista
For Dummies
978-0-471-75421-3

Available wherever books are sold. For more information or to order direct: U.S. customers visit www.dummies.com or call 1-877-762-2974.
U.K. customers visit www.wileyeurope.com or call (0) 1243 843291. Canadian customers visit www.wiley.ca or call 1-800-567-4797.